LABORATORY MANUAL
for
Conceptual Integrated Science

Paul G. Hewitt

Suzanne Lyons

John Suchocki

Jennifer Yeh

Dean Baird

with Contributions from Anne Coleman

PEARSON

Addison
Wesley

San Francisco   Boston   New York
Capetown   Hong Kong   London   Madrid   Mexico City
Montreal   Munich   Paris   Singapore   Sydney   Tokyo   Toronto   Toronto

Editor-in-Chief: Adam R.S. Black, Ph.D.
Senior Acquisitions Editor: Lothlórien Homet
Editorial Assistant: Ashley Taylor Anderson
Managing Editor: Corinne Benson
Production Superviser: Lori Newman
Manufacturing Buyer: Pam Augspurger
Director of Marketing: Christy Lawrence
Cover Designer: Richard Whitaker, Seventeenth Street Studios
Cover Photo: Photo Researchers Inc.
Project Management: Progressive Publishing Alternatives
Composition: Progressive Information Technologies
Illustrations: Progressive Publishing Alternatives
Cover and Text Printer: Bradford and Bigelow

ISBN: 0-8053-9073-1

PEARSON
Addison
Wesley

1 2 3 4 5 6 7 8 9 10 –B&B– 10 09 08 07 06
www.aw-bc.com

Table of Contents

Introduction

Many students enter this integrated science course with very few hands-on science learning experiences. The activities and experiments in this manual are designed to provide the experiences students need. Whether it's using a battery and wires to get a bulb to light, collecting carbon dioxide by water displacement, or looking through a microscope to explore the world of the unseen, these explorations are designed to put students in direct interaction with the natural world in ways that will deepen both their curiosity and their understanding. It has been suggested that we retain about 5% of what we read and about 75% of what we do, which means that the activities and experiments in this manual play an important role in the complete *Conceptual Integrated Science* program of instruction. We hope you enjoy learning by doing!

Acknowledgments

Several physics experiments and activities in this manual originated with Paul (Pablo) Robinson, author of the lab manuals for both the high school and college versions of *Conceptual Physics*. So we owe a big thank you to Pablo.

Thanks to Earl R. Feltyberger, of Nicolet High School in Glendale, Wisconsin, for the activity "Bouncy Board."

Thanks to Ted Brattstrom and Mary Graff for "Mystery Powders." For "Salt and Sand," we are grateful to Erwin W. Richter.

For contributions to nearly all the earth science labs, we are grateful to Leslie Hewitt and Bob Abrams (also a geologist). For valuable feedback and suggestions, we thank City College of San Francisco physics and geology instructor Jim Court. For other geology resources, we are grateful to the American Geological Institute and the National Association of Geology Teachers.

For astronomy activities and experiments, we are thankful to Ted Brattstrom for "Reckoning Latitude," and Forest Luke for "Tracking Mars." Both Ted and Forrest are high school teachers and amateur astronomers in Hawaii—Ted at Pearl City High School, and Forrest at Leilehua High School. For valuable feedback and advice on most of the astronomy material, we are grateful to Richard Crowe, University of Hawaii at Hilo, and to John Hubisz, North Carolina State University.

For general suggestions and feedback on all aspects of this manual, we remain indebted to mentors Charlie Spiegel and Marshall Ellenstein.

Important Notes About Safety

When performing the activities and experiments in this manual, you should always keep the safety of yourself and others in mind. Most safety rules that must be followed involve common sense. For example, if you are ever unsure of a procedure or chemical, ask your instructor, who should always be there to help.

In this manual, you will find several types of safety icons posted by selected activities. Here are the icons and what they indicate:

 **Wear approved safety goggles.** Wear goggles when working with a chemical or solution, or when heating substances.

 **Wear gloves.** Wear gloves when working with chemicals.

 **Flame/heat.** Keep combustible items such as paper towels away from open flame. Handle hot items with tongs, oven mitts, or pot holders. Do not put your hands or face over any boiling liquid. Use only heat-proof glass, and never point a heated test tube or other container at anyone. Turn off the heat source when you are finished with it.

Here are some specific safety rules that should be practiced at all times:

1. Do not eat or drink in the laboratory.

2. Maintain a clean and orderly work space. Clean up spills at once or ask for assistance in doing so.

3. Do not perform unauthorized experiments—you should always obtain permission from your instructor first. It is important that others know what you are doing and when you are doing it.

4. Do not taste any chemicals or directly breathe any chemical vapors.

5. Check all chemical labels for both name and concentration.

6. Do not grasp recently heated glassware, clamps, or other heated equipment because they remain hot for quite a while.

7. Discard all excess reagents or products in the proper waste containers.

8. If your skin comes in contact with a chemical, rinse under cold water for at least 15 minutes.

9. Do not work with flammable solvents near an open flame.

10. Assume any chemical is hazardous if you are unsure.

Master List of Supplies

Item	Activities and Experiments
acetate strips	*A Force to Be Reckoned*
acetic acid	*Smells Great!*
acetone	*Chemical Personalities*
acrylic tube	*Dropping the Ball*
aerosol room deodorizer	*In and Out*
air core solenoid	*Generator Activator*
alcohol, rubbing	*Mystery Powders, Circular Rainbows*
aluminum foil	*I'm Melting! I'm Melting!, Pinhole Camera, Solar Power I*
ammeter, DC	*Ohm, Ohm on the Range, An Open and Short Case*
ammonia cleanser	*Sensing pH*
angiosperm sample	*All Plants Are Not Created Equal*
baking soda	*Mystery Powders, Bubble Round-Up, Sensing pH*
balance, equal arm	*Spiked Water*
balance, precision	*Making Cents, Specific Heat Capacities, Upset Stomach*
ball, steel	*Rolling Stop, Dropping the Ball, Bull's Eye*
balloons	*Charging Ahead, Breathe In, Breathe Out*
barium chloride powder	*Bright Lights*
batteries, 9-volt	*Sensing pH*
batteries, D-cell	*Batteries and Bulbs, An Open and Short Case, Motor Madness*
battery, lantern (6-volt)	*Electric Magnetism*
BB shot	*Thickness of a BB Pancake*
beakers	*Specific Heat Capacities, Mystery Powders, Circular Rainbows, In and Out*
beakers, 1000-mL	*Bubble Round-Up*
beakers, 250-mL	*Chemical Personalities, Tubular Rust*
beakers, 500-mL	*Chemical Personalities*
beans	*Prey vs. Predators*
benzyl alcohol	*Smells Great!*
boiling chips	*Chemical Personalities*
borax	*Sensing pH*
boric acid	*Mystery Powders*
bottles, glass ketchup	*Pure Sweetness*
bowls	*Magnifying Microscopes*
bricks	*A Force to Be Reckoned*
brown sugar	*Pure Sweetness*
bubble solution	*Charging Ahead*
bucket, 3-gallon	*Temperature Mix*
bulb sockets, miniature	*Ohm, Ohm on the Range, Batteries and Bulbs, An Open and Short Case, Be the Battery*
bulbs, miniature	*Ohm, Ohm on the Range, Batteries and Bulbs, An Open and Short Case, Be the Battery*
Bunsen burner	*Bright Lights*
buret clamps	*Dropping the Ball, Upset Stomach*
buret stands	*Upset Stomach*
calcium chloride powder	*Bright Lights*
calcium chloride solution	*Chemical Personalities*
can, empty	*Bull's Eye*
candle	*Tuning the Senses*

Laboratory Manual for *Conceptual Integrated Science*, © 2007 Addison Wesley

cans, radiation	*Canned Heat I, Canned Heat II*
cards, 3″ × 5″	*Pinhole Image, Sunballs*
carpet strips (4 m × 50 cm)	*Rolling Stop*
celery, wilted with leaves	*In and Out*
club soda	*Sensing pH*
collar hooks	*Walking the Plank, The Weight, Motor Madness*
colored pencils	*Bright Lights*
compasses, geometric	*Over and Under, Tracking Mars*
compasses, navigational	*Electric Magnetism*
computer with motion graphing software	*Sonic Ranger, Putting the Force Before the Cart*
cooking pot	*Pure Sweetness, Sensing pH*
copper acetate	*Crystal Growth*
cornstarch	*Mystery Powders*
cotton balls	*The Amazing Senses*
coverslips	*Magnifying Microscopes, What Is It—Bacterium? Protist? Fungi?*
crossbars (short rods)	*Walking the Plank, The Weight*
crucible	*Circular Rainbows*
cupric chloride powder	*Bright Lights*
cupric sulfate pentahydrate crystals	*Chemical Personalities*
cups	*Prey vs. Predators*
cups of water	*Breathe In, Breathe Out*
cups, clear plastic	*Sensing pH*
deck of playing cards	*Ufroom Pollywoggles*
defrosting trays	*I'm Melting! I'm Melting!*
diethyl acetate (fingernail polish remover)	*Circular Rainbows*
diffraction grating	*Bright Lights*
dollar bill	*Reaction Time*
dominoes	*Chain Reaction*
dowel, wood	*Motor Madness*
drinking glass	*In and Out*
drinking straws	*Reckoning Latitude*
dynamics cart and track	*Putting the Force Before the Cart, An Uphill Climb*
eggs, raw	*Egg Toss*
elements and compounds	*Chemical Personalities*
Elodea leaves	*Magnifying Microscopes*
epsom salt	*Mystery Powders*
Erlenmeyer flasks, 250-mL	*Bubble Round-Up, Upset Stomach*
ethanol	*Oleic Acid Pancake, Circular Rainbows, Name that Recyclable*
evaporating dishes	*Chemical Personalities*
eyedropper	*Oleic Acid Pancake, Chemical Personalities, Mystery Powder, Magnifying Microscopes, What Is It—Bacterium? Protist? Fungi?*
fern with roots	*All Plants Are Not Created Equal*
food coloring	*Dance of the Molecules, Solar Power II, In and Out*
food web handouts	*Prey vs. Predators*
forceps	*Crystal Growth*
gair pins	*The Amazing Senses*
galvanometer	*Generator Activator*
garbage bags	*Egg Toss*

gas discharge tubes	*Bright Lights*
generator, hand-crank	*Be the Battery, Motor Madness, Generator Activator*
glass slides	*Crystal Growth*
gloves	*Pure Sweetness, Upset Stomach*
glue	*Muscles and Bones*
graduated cylinders, 100-mL	*Thickness of a BB Pancake, Solar Power II*
graduated cylinders, 10-mL	*Oleic Acid Pancake*
grapefruit juice	*Sensing pH*
graph paper	*Prey vs. Predators*
gymnosperm	*All PlantsAre Not Created Equal*
hardness set	*What's That Mineral?*
heat lamp and base	*Canned Heat I*
heat-proof glove or potholder	*Bright Lights*
hex nuts	*Putting the Force Before the Cart*
hot plate	*Specific Heat Capacities, Chemical Personalities, Crystal Growth*
hydrochloric acid (HCl)	*Bright Lights, Chemical Personalities, What's That Mineral?, What's That Rock?*
hypertonic solution	*Magnifying Microscopes*
ice cubes	*I'm Melting! I'm Melting!, Chemical Personalities*
index cards	*The Amazing Senses*
Internet	*Real-Life Inheritance, Investigating Evolution, Ecological Footprints*
iodine crystals	*Chemical Personalities*
iron filings	*Magnetic Personality*
isoamyl alcohol	*Smells Great!*
isobutanol	*Smells Great!*
isopentenol	*Smells Great!*
isotonic solution	*Magnifying Microscopes*
jars, baby food	*Dance of the Molecules, Pure Sweetness*
jars, glass	*Solar Power I, Indoor Clouds*
Jell-O	*In and Out*
kitchen knife	*Pure Sweetness*
leaf assortment	*All Plants Are Not Created Equal*
leaf identification chart	*All Plants Are Not Created Equal*
lens, 25-mm converging	*Pinhole Camera*
library access	*Real-Life Inheritance*
lightbulbs, clear 100 watt	*Solar Power I*
liter container	*Temperature Mix*
lithium chloride powder	*Bright Lights*
luminol crystals	*Chemical Personalities*
lycopodium powder	*Oleic Acid Pancake*
magazines, muscle	*Muscles and Bones*
magnet, neodymium	*Dropping the Ball*
magnetic field projectual	*Magnetic Personality*
magnets, bar	*A Force to be Reckoned, Magnetic Personality, Motor Madness, Generator Activator*
magnifying glasses	*All Plants Are Not Created Equal*
markers	*Muscles and Bones*
mass blocks	*Putting the Force Before the Cart*

Laboratory Manual for *Conceptual Integrated Science*, © 2007 Addison Wesley

mass hanger	*The Weight*
masses, hooked	*Bouncy Board*
masses, slotted	*Walking the Plank, The Weight*
matches	*Tuning the Senses, Charging Ahead*
measuring cup	*Indoor Clouds*
metal specimens	*Specific Heat Capacities*
meterstick	*Go! Go! Go!, Walking the Plank, Bouncy Board, An Uphill Climb, Rolling Stop, Dropping the Ball, Pinhole Image, Solar Power I, Solar Power II, Sunballs*
methanol	*Chemical Personalities, Circular Rainbows, Smells Great!*
micrometer	*Thickness of a BB Pancake*
microscopes	*Crystal Growth, Magnifying Microscopes, What Is It?*
microspatula	*Chemical Personalities*
mineral collection	*What's That Mineral?*
mirror, full length	*Mirror, Mirror, on the Wall*
molecular modeling kit	*Molecules by Acme*
mortar and pestle	*Upset Stomach*
moss sample	*All plants Are Not Created Equal*
motion sensor	*Sonic Ranger, Putting the Force Before the Cart*
nails (short)	*Spiked Water*
newspaper	*Magnifying Microscopes*
n-propanol	*Smells Great!*
nuts, 1/2 inch	*Sugar Soft*
octanol	*Smells Great!*
oleic acid	*Oleic Acid Pancake*
open space	*Prey vs. Predators*
paint, flat black	*Solar Power I*
paper	*Muscles and Bones*
paper bags	*Breathe In, Breathe Out*
paper clips	*Putting the Force Before the Cart, Understanding Darwin*
paper towels	*Spiked Water, Canned Heat I, Canned Heat II, I'm Melting! I'm Melting!*
paper, butcher	*Go! Go! Go!*
paper, circular filter	*Circular Rainbows*
paper, graph	*Making Cents, Go! Go! Go!, The Weight, Canned Heat I, Canned Heat II, Ohm, Ohm on the Range, Sugar Soft, Tracking Mars*
pencils, colored	*Over and Under*
pencils, regular	*Keep Pumping*
pennies	*Making Cents*
pens, black felt-tip	*I'm Melting! I'm Melting!, Circular Rainbows*
petri dishes	*Crystal Growth*
phenolphthalein	*Mystery Powders*
photogate timers	*Dropping the Ball*
pie tins, small	*Charging Ahead*
pineapple slices, fresh	*In and Out*
pins, straight	*Pinhole Image*
pipets, 9-inch plastic	*Sugar Soft, Upset Stomach*
pith balls	*A Force to be Reckoned*
place to run	*Keep Pumping*
plaster of Paris	*Mystery Powders*
plastic spoons	*Understanding Darwin*

plastic wrap	*Solar Power II*
playing cubes (or painted sugar cubes)	*Get a Half-life!*
plumb line and bob	*Reckoning Latitude*
pond water, fresh	*What Is It—Bacterium? Protist? Fungi?*
potassium aluminum sulfate (alum)	*Crystal Growth*
potassium chloride powder	*Bright Lights*
potato	*Magnifying Microscopes*
power resistors	*Ohm, Ohm on the Range*
power supply, variable DC (0–6 V)	*Ohm, Ohm on the Range*
printer	*Ecological Footprints*
propionic acid	*Smells Great!*
protractor	*An Uphill Climb, Over and Under, Walking on Water, Reckoning Latitude, Tracking Mars*
PTC paper	*Real-Life Inheritance*
pulley	*Putting the Force Before the Cart*
pushpins	*Reckoning Latitude*
ramp (2 meters)	*Rolling Stop*
recyclable plastics	*Name That Recyclable*
red cabbage	*Sensing pH*
ring clamp	*Electric Magnetism*
ring stand	*Bright Lights, Chemical Personalities, Bubble Round-Up, Tubular Rust*
rock samples	*What's That Rock?*
rod clamps	*Walking the Plank, An Uphill Climb, The Weight, Motor Madness*
rubber bands	*Motor Madness, Tubular Rust, Solar Power II*
rubber stoppers, drilled	*Bubble Round-Up, Solar Power I*
ruler, centimeter	*Reaction Time, Mirror, Mirror, on the Wall, Thickness of a BB Pancake, Sugar Soft, Tubular Rust, Top This, Walking on Water, The Amazing Senses, Tracking Mars*
safety goggles	*Egg Toss, Mystery Powders, Salt and Sand, Bubble Round-Up, Pure Sweetness, Sensing pH, Upset Stomach*
salicylic acid	*Smells Great!*
salt, table	*Mystery Powders, Salt and Sand, Sensing pH, Magnifying Microscopes*
saltwater	*Magnifying Microscopes*
sand	*Salt and Sand*
scissors	*Magnifying Microscopes, Muscles and Bones*
shoebox	*Pinhole Camera*
silk cloth square	*A Force to Be Reckoned*
skeleton, model	*Muscles and Bones*
slides	*Magnifying Microscopes, What Is It—Bacterium? Protist? Fungi?*
slides, prepared for viewing	*Magnifying Microscopes*
sodium carbonate solution	*Chemical Personalities*
sodium chloride	*Crystal Growth*
sodium chloride powder	*Bright Lights*
sodium chloride solution	*Chemical Personalities*
sodium hydroxide	*Mystery Powders*
sodium nitrate	*Crystal Growth*
sodium perborate crystals	*Chemical Personalities*
sodium sulfate solution	*Chemical Personalities*
solder, lead-free	*Motor Madness*

Laboratory Manual for *Conceptual Integrated Science*, © 2007 Addison Wesley

spatula	*Mystery Powders*
spectroscope	*Bright Lights*
spring scales	*Walking the Plank, An Uphill Climb, The Weight*
steel wool	*Chemical Personalities, Tubular Rust*
stereo audio device	*Sound Off*
stirring rod, glass	*Chemical Personalities*
stopwatch	*Go! Go! Go!, Chain Reaction, Keep Pumping, In and Out*
strainer	*Sensing pH*
straws	*Breathe In, Breathe Out*
streak plates	*What's That Mineral?*
string	*Walking the Plank, Putting the Force Before the Cart, Ellipses*
strobe light, variable frequency	*Slow-Motion Wobbler*
strontium chloride powder	*Bright Lights*
styrofoam bowls	*Charging Ahead*
styrofoam cups	*Temperature Mix, Spiked Water, Specific Heat Capacities, Solar Power II*
styrofoam plates	*I'm Melting! I'm Melting!*
sucrose crystals	*Chemical Personalities*
sugar, table	*Mystery Powders, Sugar Soft*
support rod	*Walking the Plank, An Uphill Climb, Dropping the Ball, The Weight, Electric Magnetism, Motor Madness*
switch, double pole double throw	*Sound Off*
table clamps	*Walking the Plank, An Uphill Climb, Dropping the Ball, The Weight*
tape measures	*The Amazing Senses, Breathe In, Breathe Out*
tape, clear	*Solar Power I, Reckoning Latitude, In and Out*
tape, masking	*Go! Go! Go!, Egg Toss, Mirror, Mirror, on the Wall, Pinhole Camera*
test tube clamps	*Chemical Personalities, Tubular Rust*
test tube, large	*Chemical Personalities*
test tube, medium with stopper	*Chemical Personalities*
test tubes (15 cm)	*Mystery Powders, Tubular Rust*
textbooks, old	*Muscles and Bones*
thermometer (Celsius)	*Spiked Water, Specific Heat Capacities, Canned Heat I, Canned Heat II, Chemical Personalities, Solar Power I, Solar Power II, In and Out*
thumbtacks	*Ellipses*
thymol	*Crystal Growth*
tincture of iodine	*Mystery Powders*
toilet bowl cleaner	*Sensing pH*
tongs	*Specific Heat Capacities*
topographic maps	*Top This*
toy car, constant velocity	*Go! Go! Go!*
tracing paper	*Pinhole Camera*
tray	*Thickness of a BB Pancake, Oleic Acid Pancake, Indoor Clouds*
tubing	*Bubble Round-Up*
tuning forks, low frequency	*Slow-Motion Wobbler*
twine	*Bouncy Board*
Van de Graaff generator	*Charging Ahead*
vinyl strips	*A Force to Be Reckoned*
voltmeter, DC	*Ohm, Ohm on the Range*

washing soda   *Mystery Powders, Sensing pH*
water   *In and Out*
weigh dishes   *Upset Stomach*
well-plates   *Chemical Personalities, Upset Stomach*
white chalk   *Mystery Powders*
white vinegar   *Mystery Powders, Bubble Round-Up, Circular Rainbows, Sensing pH, Tubular Rust*
wires, connecting   *Ohm, Ohm on the Range, Batteries and Bulbs, An Open and Short Case, Be the Battery, Electric Magnetism, Motor Madness, Generator Activator*
wood block   *Putting the Force Before the Cart, Motor Madness*
wool cloth squares   *A Force to Be Reckoned*
word-processing program *Ecological Footprints*

Name _____ Date _____

CONCEPTUAL INTEGRATED SCIENCE	Activity

About Science

Tuning the Senses

Purpose
In this activity, you will tune your senses of sight and sound.

Scientists' original source of information about the universe comes from personal observations. This leads to questioning reasons and causes. A scientist notices something, asks questions, and then tries to answer them. By this definition, can't we all be scientists?

Required Equipment and Supplies
notebook, pen or pencil, candle, matches, and patience

Discussion
Galileo wrote, "In questions of science the authority of a thousand is not worth more than the humble observation and reasoning of a single individual." We'll do two simple activities: the first to tune our hearing, the second to tune our seeing.

Procedure
Perform the following activities on your own and then answer the questions on the following page.

Activity 1: Audition of the Environment
Go outside and find a comfortable place to sit. Listen to your environment for 10 minutes. Write down the sounds you hear. You might find it helpful to close your eyes during this activity—don't let the sights of your surroundings distract you from observing its sounds.

Activity 2: Observation of a Burning Candle
Remain absolutely quiet while observing an unlit candle for 2 minutes. Record your observations.

Light the candle, observe it for 2 minutes, then record your observations.

After you have exhausted all observations possible, extinguish the candle.

Observe the extinguished candle while recording more observations.

What other activities can you invent to tune your senses?

Be creative and tune your senses every day!

Questions for Activity 1: Audition of the Environment

1. What was the quietest sound you heard? What was the loudest?

_____

2. Which sounds had a relatively high pitch (a light breeze, for example)?

_____

3. Which sounds had a relatively low pitch (a car engine, for example)?

_____

4. How many sounds were natural, and how many were man made?

_____

5. Did you identify sounds by their sources, or did you spell them out phonetically?

_____

6. Are there any continuous sounds *right now* (like that of an air conditioner) that you have simply tuned out? What are they?

_____

7. How might you describe a sound to someone who was unfamiliar with its source (for example, the sound of a car to someone who had never heard of a car before)?

_____

Questions for Activity 2: Observation of a Burning Candle

8. Did you detect any odors? If so, describe them.

_____

9. What color was the molten wax?

_____

10. Did you note the time and date of your candle observations?

_____

11. Was it hotter 6 inches above the flame or 6 inches to the side of the flame (or was it equally hot in both places)?

_____

12. Describe the patterns in the smoke before and after the flame was extinguished.

_____

13. Was the color transition of blue to yellow in the flame gradual or abrupt?

_____

14. Did you describe the candle by drawing a picture of it?

_____

CONCEPTUAL INTEGRATED SCIENCE	Activity

About Science: The Scientific Method

Making Cents

Purpose
In this activity, you will investigate the relationship between the mass of a penny and its age.

Required Equipment and Supplies
10 pennies per student
balances
graph paper

Discussion
The scientific method is an effective way of gaining, organizing, and applying new knowledge. A common form of the method is essentially as follows:

1. Recognize a problem.

2. Make an educated guess—a **hypothesis.**

3. Predict the consequences of the hypothesis.

4. Perform experiments to test predictions. If necessary, modify the hypothesis in light of experimental results. Perform more experiments.

5. Formulate the simplest general rule that organizes the three main ingredients—hypothesis, prediction, and experimental outcome.

Procedure
Step 1: Propose a hypothesis to the following question (problem): What effect does aging have on a penny's mass?

Step 2: Based on your hypothesis, predict the general form of a graph that plots the mass of a penny (*y*-coordinate) relative to its age (*x*-coordinate).

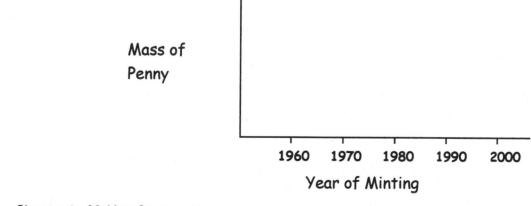

Step 3: Using a balance, measure the mass of at least 10 individual pennies minted in different years. Enter the mass in grams of each penny relative to the year it was minted (Table 1).

Table 1

Mass of Penny:										
Year Minted:										

Step 4: Pool your data with that of all other students on the class chalkboard, and create you own graph using all this data. Show the mass of each penny in grams on the *y*-coordinate and the year the penny was minted on the *x*-coordinate. Alternatively, data may be entered into a computer program that will plot the graph for you.

Step 5: Are the experimental results consistent with your hypothesis? If not, propose a new hypothesis.

Step 6: If you have formed a new hypothesis, what additional measurements might you take to support this new hypothesis? Perform these measurements, and record your results and observations here:

Summing Up

1. What conclusion can you draw from the results of your experimental data?

_____

_____

2. What effect might aging have on the mass of a nickel, a dime, and a quarter?

_____

_____

3. Would using a balance that was many times more sensitive have made a difference in your conclusion about the effect of aging on a penny? Briefly explain.

_____

_____

4. What improvements might you expect in your graph if only one student had done all the weighing on a single balance?

_____

_____

Laboratory Manual for *Conceptual Integrated Science,* © 2007 Addison Wesley

| **CONCEPTUAL INTEGRATED SCIENCE** | **Experiment** |

Describing Motion: Speed and Velocity

Go! Go! Go!

Purpose

In this experiment, you will plot a graph that represents the motion of an object.

Required Equipment and Supplies

constant velocity toy car
butcher paper or continuous (unperforated) paper towel
access to tape
stopwatch
meterstick
graph paper

Discussion

Sometimes two quantities are related to each other, and the relationship is easy to see. Sometimes the relationship is harder to see. In either case, a graph of the two quantities often reveals the nature of the relationship. In this experiment, we will plot a graph that represents the motion of a real object.

Procedure

We are going to observe the motion of the toy car. By keeping track of its position relative to time, we will be able to make a graph representing its motion. To do this, we will let the car run along a length of butcher paper. At 1-second intervals, we will mark the position of the car. This will result in several ordered pairs of data—positions at corresponding times. We can then plot these ordered pairs to make a graph representing the motion of the car.

Step 1: Fasten the butcher paper to the top of your table. It should be as flat as possible—no hills or ripples.

Step 2: If the speed of the toy car is adjustable, set it to the slow setting.

Step 3: Aim the car so that it will run the length of your table. Turn it on, and give it a few trial runs to check the alignment.

Step 4: Practice using the stopwatch. For this experiment, the stopwatch operator needs to call out something like, "Go!" at each 1-second interval. Try it to get a sense of the 1-second rhythm.

Step 5: Practice the task.
  a. Let the car drive across the length of the butcher paper.
  b. Soon after it starts, the stopwatch operator will start the stopwatch and say, "Go!"
  c. Another person in the group should practice marking the location of the front or back of the car on the butcher paper every time the watch operator says, "Go!" For the practice run, simply touch the eraser of the pencil to the butcher paper at the appropriate points.
  d. The watch operator continues to call out "Go!" once each second, and the marker continues to practice marking the location of the car until the car reaches the end of the butcher paper or table. Take care to keep the car from running off the table!

Step 6: Perform the task.

    a. Let the car drive across the length of the butcher paper.

    b. Soon after it starts, the stopwatch operator will start the stopwatch and say "Go!"

    c. Another person in the group will mark the location of the front or back of the car on the butcher paper every time the watch operator says, "Go!" **No marks are to be made on the paper until the car is moving.**

    d. The watch operator continues to call out, "Go!" once each second, and the marker continues to mark the location of the car until the car reaches the end of the butcher paper or table. Take care to keep the car from running off the table!

Step 7: Label the marked points. The first mark is labeled "0," the second is labeled "1," the third is "2," and so on. These labels represent the time at which the mark was made.

Step 8: Measure the distances—in centimeters (cm)—of each point from the point labeled "0." (The "0" point is 0 cm from itself.) Record the distances on the data table. Don't worry if you don't have as many data points as there are spaces available on the data table.

Data Table

Time t (s)	0	1	2	3	4			
Position x (cm)	0							

Step 9: Make a plot of position vs. time on the graph paper. Title the graph "Position vs. Time." Make the horizontal axis *time* and the vertical axis *position*. Label the horizontal axis with the quantity's symbol and the units of measure: "*t*(s)". Label the vertical axis in a similar manner. Make a scale on both axes starting at 0 and extending far enough so that all your data will fit within the graph. Don't necessarily make each square equal to 1 second or 1 cm. Make the scale so the data will fill the maximum area of the graph.

We could just as easily make a graph of time vs. position. But we prefer position vs. time for a few reasons. In this experiment, time is what we call an "independent variable." That is, no matter how fast or slow our car was, we always marked its position at equal time intervals. *We* were in charge of the time intervals; the *car* was "in charge" of the change in position it made in each interval. But the change in position of the car in each interval depended on the time interval we chose. So we call position the "dependent variable." We generally arrange a graph so that the horizontal axis represents the independent variable, and the vertical axis represents the dependent variable. Also, the slope of a position vs. time graph tells us more than the slope of a time vs. position graph, as we will see later.

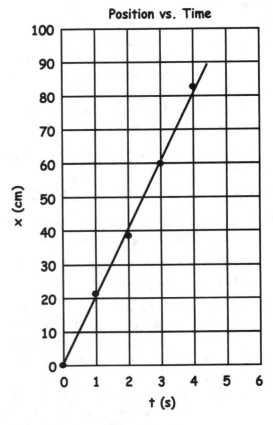

Figure 1

    Laboratory Manual for *Conceptual Integrated Science,* © 2007 Addison Wesley

Step 10: Draw a line of best fit. In this case, the line of best fit should be a single, straight line. Use a ruler or straight edge; place it across your data points so that your line will pass as close as possible to all your data points. The line may pass above some points and below others. Don't simply draw a line connecting the first point to the last point. An example is shown in Figure 1.

Step 11: Determine the slope of the line. Slope is often referred to as "rise over run." To determine the slope of your line, proceed as follows.

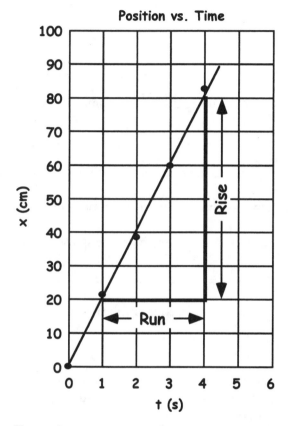

Figure 2

a. Pick two convenient points on your line. They should be pretty far from each other. Convenient points are those that intersect grid lines on the graph paper.

b. Extend a horizontal line to the right of the lower convenient point, and extend a vertical line downward from the upper convenient point until you have a triangle as shown in Figure 2. It will be a right triangle, because the horizontal and vertical lines meet at a right angle.

c. Find the length of the horizontal line on your graph. This is the "run." Don't use a ruler; the length must be expressed in units of the quantity on the horizontal axis. In this case, seconds of time.

Run: _____ s

d. Measure the length of the vertical line. This is the "rise." Don't use a ruler; the length must be expressed in units of the quantity on the vertical axis, in this case, centimeters of distance.

Rise: _____ cm

e. Calculate the slope by dividing the rise by the run. Show your calculation below.

Slope: _____ cm/s

Summing Up

1. Suppose a faster car were used in this experiment. What would have been different about

a. the distance between the marks on the butcher paper?

_____

b. the number of seconds the car would have spent on the butcher paper before reaching the edge?

_____

c. the resulting distance vs. time graph? (How would the slope have been different?)

_____

2. Add a line to your graph that represents a faster car. Label it appropriately.

3. Suppose a slower car were used in this experiment. What would have been different about

   a. the distance between the marks on the butcher paper?

   _____

   b. the number of seconds the car would have spent on the butcher paper before reaching the edge?

   _____

   c. the resulting distance vs. time graph? (How would the slope have been different?)

   _____

4. Add a line to your graph that represents a slower car. Label it appropriately.

5. Suppose the car's battery ran out during the run so that the car slowly came to a stop.

   a. What would happen to the space between the marks as the car slowed down?

   _____

   b. Add a line to your graph that represents a car slowing down. Label it appropriately.

6. What motions do these graphs represent? In other words, what was the car doing to generate these motion graphs?

Line A. _____

_____

_____

_____

Line B. _____

_____

_____

_____

Position vs. Time

CONCEPTUAL INTEGRATED SCIENCE	Activity

Describing Motion: Speed and Velocity

Sonic Ranger

Purpose
In this activity, you will use graphs to investigate motion. The graphs will represent your own motion and will be drawn by the computer as you move.

Required Equipment and Supplies
sonic ranging device with appropriate equipment and software
computer

Discussion
Graphs can be used to represent motion. For example, if you track the position of an object as time goes by, you can make a plot of position vs. time. In this activity, the sonic ranger will track your position, and the computer will draw a position vs. time graph of your motion. The sonic ranger sends out a pulse of high frequency sound and then listens for the echo. By keeping track of how much time goes by between each pulse and corresponding echo, the ranger determines how far you are from it. (Bats use this technique to navigate in the dark.) By continually sending pulses and listening for echoes, the sonic ranger tracks your position over a period of time. This information is fed to the computer, and the software generates a position vs. time graph.

Procedure
Your instructor will provide a computer with a sonic ranging program installed. Check to see that the sonic ranger is properly connected and operating reliably. Position the sonic ranger so that its beam is about chest high and aimed horizontally. (Note: Sometimes these devices do not operate reliably on top of computer monitors.)

The sonic ranger should be set to "long range" mode. The computer should be set to graph position vs. time. Initiate the sonic ranger and note how close and how far you can get before the readings become unreliable.

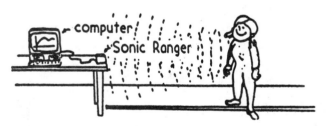

Part A: Move to Match the Graph
Generate real-time graphs of each motion depicted on pg. 10 and write a description of each. *Do not use the term "acceleration" in any of your descriptions.* Instead, use terms and phrases such as, "rest," "constant speed," "speed up," "slow down," "toward the sensor," and "away from the sensor."

Study each graph on pg. 10. When you are ready, initiate the sonic ranger and move so that your motion generates a similar graph. Then describe the motion in words.

Example:

Position vs. Time Graph

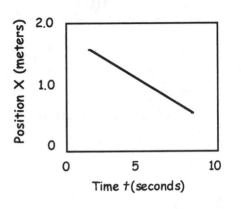

Description
Move toward the sensor at constant speed.

Make sure each person in the group can move to match this graph before moving on to the next graph.

1.

Position vs. Time Graph

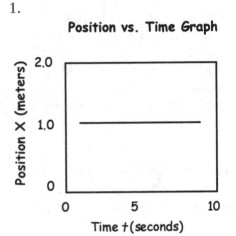

Description

2.

Position vs. Time Graph

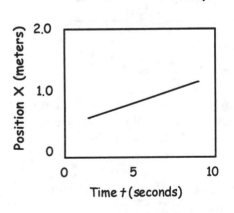

Description

Laboratory Manual for *Conceptual Integrated Science,* © 2007 Addison Wesley

3.

Position vs. Time Graph

Description

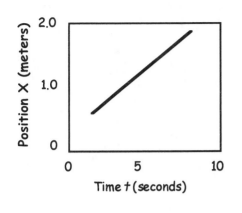

4.

Position vs. Time Graph

Description

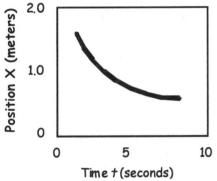

Part B: Move to Match the Words

Walk to match each description of motion. Draw the resulting position vs. time graph.

5.

Description

Move toward the sensor at constant speed, stop and remain still for a second, then walk away from the sensor with constant speed.

Position vs. Time Graph

6.

Description

Move toward the sensor with decreasing speed, then just as you come to rest, move away from the detector with increasing speed.

Position vs. Time Graph

7.

Description	Position vs. Time Graph

Move away from the sensor with decreasing speed until you come to a stop. Then move toward the sensor with decreasing speed until you come to a stop.

Summing Up

1. How does the graph show the difference between forward motion and backward motion?

_____

2. How does the graph show the difference between slow motion and fast motion?

_____

3. Study the graph of position vs. time shown below.

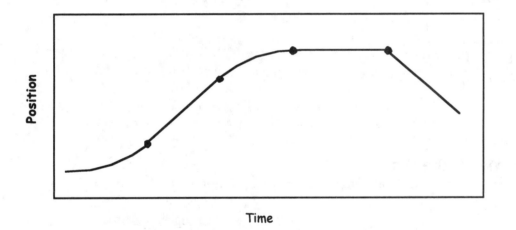

Label the sections of the graph showing where the object is
- at rest.
- moving forward at constant speed.
- moving backward at constant speed.
- speeding up.
- slowing down.

CONCEPTUAL INTEGRATED SCIENCE	Activity

Describing Motion: The Equilibrium Rule
Walking the Plank

Purpose
In this activity, you will measure and interpret the forces acting on an object when the object is in equilibrium.

Required Equipment and Supplies
meterstick
2 table clamps
2 support rods
2 crossbars (short rods)
2 rod clamps
2 collar hooks
2 spring scales (5- or 10-newton capacity)
slotted masses [two 200 gram (g) and one 500 g]
2 20 centimeter (cm) lengths of string
small spirit level (optional)

Discussion
Consider two sign painters who work on a scaffold (a plank of wood suspended by ropes at both ends). They might wonder about the tension in the ropes that support their platform. They are in a state of equilibrium, but how do the forces relate to one another? Their weights are downward forces, and the tensions in the ropes suspending the scaffold are upward forces. While the weights of the painters never change, the tensions in the ropes seem to change when the painters move along the platform. In this activity, you will arrange a platform similar to a painter's scaffold. You will measure the forces acting on the scaffold when it is in various arrangements. You will interpret the forces to determine the condition of equilibrium.

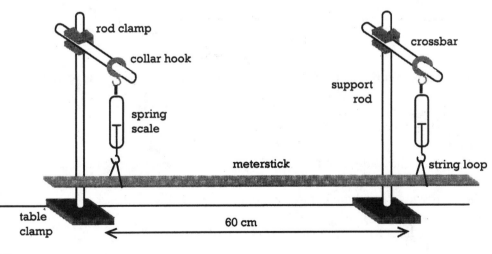

Figure 1

Procedure

Step 1: Calibrate both spring scales so that when held vertically and carrying no load, each reads zero.

Step 2: Arrange the apparatus as shown in Figure 1.

    a. Position the table clamps so that the support rods are about 60 cm apart.

    b. Attach the crossbars to the support rods using the clamps. Hang a spring scale from each of the crossbars using the collar hooks.

    c. Tie the ends of one 20-cm length of string together to create a loop. Hang the loop from one of the spring scales. Repeat for the other spring scale.

    d. Suspend the meterstick (centimeter scale up) from the loops of string. Balance the arrangement so that the 50-cm mark is centered between the string loops and the meterstick is level. (Use a spirit level—if one is available—to check the meterstick.) This structure is a model of our painters' scaffold.

    e. Adjust the meterstick so that the readings on both spring scales are the same (or very nearly the same). Move the meterstick left or right or adjust the level if necessary.

Step 3: Record the readings on both scales.

Reading on left scale:_____ Reading on right scale:_____

1. Add those readings and record the result. This is the total weight of the meterstick and string loops.

_____

2. Complete the diagram of the meterstick with the forces acting on it. The force L is the upward force on the left, R is the force on the right, and W is the downward force of weight

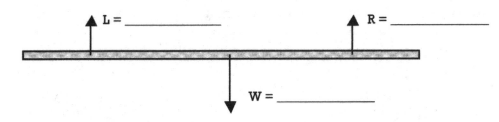

3. What is the net force on the meterstick? The net force is the sum of the forces, taking direction into account.

_____

Step 4: Carefully place one 200-g mass at the 40-cm mark while carefully placing the other 200-g mass at the 60-cm mark. (These represent our painters; take care so they don't fall!) Aim the slots of the slotted masses toward either end of the meterstick (0 or 100 cm).

Step 5: Record the readings for both scales.

Reading of left scale:_____ Reading of right scale:_____

4. What is the total weight of the meterstick, string loops, and masses?

_____

5. Sketch a diagram of the meterstick with all the forces acting on it. Include the numerical values of each force in your diagram.

6. What is the net force on the meterstick?

_____

Step 6: Move the mass at the 40-cm mark to the 70-cm mark. Keep the other mass at the 60-cm mark. The structure is still in equilibrium, even though the load is not evenly distributed.

Step 7: Record the readings for both scales.

Reading of left scale:_____   Reading of right scale:_____

7. What is the total weight of the meterstick, string loops, and masses?

_____

8. Sketch a diagram of the meterstick with all the forces acting on it. Include the numerical values of each force in your diagram.

9. What is the net force on the meterstick?

_____

10. Review your findings so far. What would you say is the condition for equilibrium, a condition that was met in all the arrangements investigated so far?

_____

_____

_____

Step 8: Suppose two painters with different weights used a scaffold. Simulate this by using a 500-g mass and a 200-g mass. Carefully stack the two masses at the 50-cm mark and read the scales.

Reading of left scale:_____   Reading of right scale:_____

Step 9: Carefully place the 200-g mass at the 60-cm mark and the 500-g mass at the 40-cm mark, but do not read the scales yet.

11. What will the scale readings add to?

_____

Step 10: Read **only** the left scale and record the reading.

Reading on left scale:_____

12. Predict the approximate value of the reading on the right scale and record your prediction.

Prediction on right scale:_____

Step 11: Read the right scale and record the reading.

Reading on right scale:_____

13. How did the reading compare to your prediction?

_____

Step 12: Move the 200-g mass until both spring scales have the same reading. Record the location of the 200-g mass.

Position of the 200-g mass:_____

14. The 500-g mass is 10 cm from the center of the meterstick. How far is the 200-g mass from the center of the meterstick?

_____

Summing Up

1. Can the meterstick platform be in equilibrium if the two upward support forces are equal to each other? If so, give an example from your observations.

_____

_____

_____

2. Can the meterstick platform be in equilibrium if the two upward support forces are unequal? If so, give an example from your observations.

_____

_____

_____

3. Would the platform be in equilibrium if a 500-g mass were at the 30-cm mark and a 200-g mass were at the
60-cm mark? Explain.

_____

_____

_____

4. Suppose the 500-g mass were placed at the 30-cm mark. Where could you place the 200-g mass so that both spring scales would have the same reading? Explain your answer.

_____

_____

_____

5. Could you use the same masses to get both scales to have the same reading if the 500-g mass were placed at the 20-cm mark? If so, where should the 200-g mass be placed? If not, why not?

_____

_____

_____

Name _____ Date _____

| CONCEPTUAL INTEGRATED SCIENCE | Activity |

Newton's Laws of Motion: Newton's Second Law of Motion

Putting the Force Before the Cart

Purpose
In this activity, you will observe the motion of a variety of objects under a variety of conditions. You will interpret your observations to learn the relationship between force, mass, and acceleration.

Required Equipment and Supplies
dynamics cart and track
mass blocks
string [about 1 meter (m)]
pulley
wood block
paper clip
4 hex nuts (or equivalent)
computer with motion graphing software
motion sensor and interface device

Discussion
Some of the most fundamental laws of motion eluded the best minds in science for centuries. One reason for this is that when objects are pushed or pulled, there are usually several forces acting at once. To understand the nature of force and motion, it is necessary to observe the effect of a single, unbalanced force acting on an object. In this activity, you will do just that. You will also vary the amount of force acting on the object, and you will vary the mass of the object being acted upon. Careful observations will lead you to an understanding of the relationship between force, mass, and acceleration.

Procedure
Step 1: Connect the motion sensor to the computer (using the interface device). If the motion sensor has a range selector, choose the "short range" setting. Activate the motion graphing software.

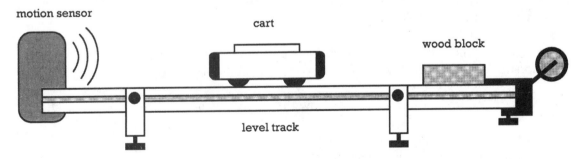

Figure 1

Step 2: Arrange the apparatus as shown in Figure 1.
  a. Make sure the track is level. The cart should be able to coast equally in either direction along the track. If the cart "prefers" to roll in one direction, adjust the track accordingly.

  b. The pulley clamp should be secure on the track, and the pulley should be able to spin freely.

c. Set the wood block on the track (or employ some other stopping mechanism) so the cart cannot roll into the pulley.

d. Arrange the string so that it is attached to the cart at one end and the paper clip at the other end, as shown in Figure 2. The length of the string is such that when the cart is stopped at the wood block, the paper clip does not touch the ground. If the paper clip touches the ground, shorten the string.

Step 3: Test the sensor and software.

a. Place the cart near the middle of the track.

b. Activate the motion sensor and the graphing software.

c. Move the cart back and forth with your hand. The computer should show a graph that corresponds to the motion of the cart. If it does not, adjust the aim of the sensor and check the connecting wires. If the problems persist, ask your instructor for assistance.

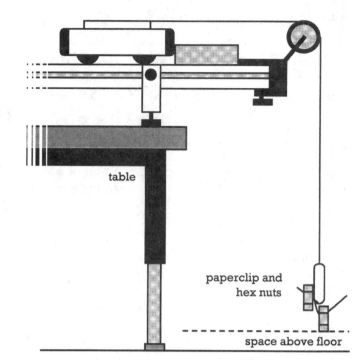

Figure 2

Part A: Vary the Force

Step 4: Check to see that there are two hex nuts attached to the paper clip. Pull the cart back so the paper clip is just below the pulley wheel. Make note of the cart's starting position on the track.

Step 5: Clear the computer of any previous trials and activate the motion sensor.

Step 6: When the motion sensor begins sampling, release the cart and allow it to move along the track until it is stopped by the wood block.

Step 7: Deactivate the motion sensor. If something went wrong during the trial, simply reset the cart, string, and software, and repeat the trial so that you have a reliable result.

1. Sketch the graph in the space to the right. Show the smooth, general pattern; neglect insignificant data point spikes and glitches. Show only the portion of the graph that corresponds to when the cart was moving. How does this graph of accelerated motion differ from a graph of uniform motion (motion having constant velocity)?

_____

_____

_____

_____

2. What do you think will happen to the acceleration if twice as much force is used to pull the cart?

_____

_____

Step 8: Add the two remaining hex nuts to the paper clip (for a total of four). This will double the force pulling the cart.

Step 9: Set the cart in place for a second trial, starting from the same position on the track. Prepare the software to add a second trial to the one already recorded. Activate the sensor. When the sensor begins sampling, release the cart. When the cart is stopped, deactivate the sensor.

3. How does the acceleration caused by the doubled force compare to the original acceleration (from Step 6)? Did your observation confirm or contradict your prediction?

_____

_____

Part B: Vary the Mass
Step 10: Determine the mass of your cart and record it here: _____

Step 11: Clear the computer of any previous trials.

Step 12: Attach four hex nuts to the paper clip. Set the cart in place. Activate the sensor. When the sensor begins sampling, release the cart. When the cart is stopped, deactivate the sensor.

Step 13: Add a mass block or blocks to the cart so that the mass is doubled. For example, if the cart has a mass of 500 g, add 500 g of mass blocks to it. Do not change the hex nut configuration.

4. What do you think will happen to the acceleration if the same force is used to pull a cart having twice as much mass?

_____

_____

Step 14: Set the cart in place for a second trial. Prepare the software to add a second trial to the one already recorded. Activate the sensor. When the clicking begins, release the cart. When the cart is stopped, deactivate the sensor.

5. How does the acceleration of the doubled mass compare to the original acceleration (from Step 12)? Did your observation confirm or contradict your prediction?

_____

_____

Summing Up

1. How does the acceleration of the cart depend on the force pulling it?

_____ Greater force results in greater acceleration. In other words, acceleration is directly proportional to force.

_____ Greater force results in lesser acceleration. In other words, acceleration is inversely proportional to force.

_____ Greater force results in the same acceleration. In other words, acceleration is independent of force.

2. How does the acceleration of the cart depend on the mass of the cart?

_____ Greater cart mass results in greater acceleration. In other words, acceleration is directly proportional to mass.

_____ Greater cart mass results in lesser acceleration. In other words, acceleration is inversely proportional to mass.

_____ Greater cart mass results in the same acceleration. In other words, acceleration is independent of mass.

3. Complete the statement:

The acceleration of an object is _____ proportional to the net force

acting on it and _____ proportional to the mass of the object.

4. Which mathematical expression is most consistent with your observations?

a. $a = F \cdot m$ b. $a = \frac{F}{m}$ c. $a = \frac{m}{F}$

5. Examine the position vs. time graphs plotted in the diagram to the right. Suppose Plot B represented an empty cart pulled by two hex nuts. If the mass of the cart were doubled and four hex nuts were used to pull the cart, which plot would best represent the result? Explain.

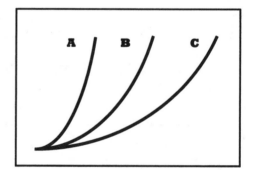

_____

_____

_____

_____

CONCEPTUAL INTEGRATED SCIENCE	Activity

Momentum and Energy: Impulse–Momentum Relationship

Egg Toss

Purpose
In this activity, you will investigate the effect that stopping time has on stopping force when momentum changes.

Required Equipment and Supplies
raw egg
garbage bags (1–2)
masking tape
safety glasses or goggles
a playing field (soccer, football, baseball, softball, etc.)
trundle wheel or long tape measure (optional)

Procedure
One person in the group is going to throw a raw egg to another person in the group. The second person must catch the egg without letting it break. When the thrower and catcher are close to each other, the task is fairly simple. As the distance increases, the task becomes more difficult.

Step 1: Prepare for the activity by choosing a thrower and a catcher. The thrower will throw the egg to the catcher. The catcher will catch the egg. Both thrower and catcher must use only their bare hands to handle the egg.

Step 2: The catcher must wear the safety goggles and a plastic poncho. Use the garbage bag and masking tape to construct a plastic poncho. When the egg breaks, it can make a mess. Make sure the poncho covers any part of your clothes that should be protected from raw egg white and yolk.

Step 3: Go to the field and line up according to your instructor's directions. The throwers and catchers should be facing each other and should start about 3 meters (m) apart from each other. Throwers should also be 2 to 3 m from each other.

Step 4: When the instructor gives the signal, the thrower throws the egg to the catcher. If the catcher catches the egg and the egg remains intact, the group may proceed to the next toss. If the egg breaks, step on the remains and grind them into the ground. If a trundle wheel or long tape measure is available, use it to determine the distance between the thrower and catcher when the egg broke.

The thrower (or someone else in the group) must retrieve the egg from the catcher. The catchers should move back 3 m, and the throwers should return to their original throwing line.

Step 5: Repeat Step 4 until the last group breaks their egg.

Summing Up

1. How far did your thrower and catcher get from each other before the egg broke? What was the longest distance achieved in the class? (Use an estimate if you didn't measure it.)

_____

_____

2. What was the trick to making a successful catch? What does this have to do with stopping time?

_____

_____

3. Compare a sudden-stop catch with a gradual-stop catch.

   a. In which case is the mass of the egg greater? Or is it the same either way?

   _____

   b. In which case is the change in velocity of the egg greater? Or is it the same either way? (Be careful!)

   _____

   c. In which case is the change in momentum ($m\Delta v$) of the egg greater? Or is it the same either way?

   _____

   d. In which case is the stopping time greater? Or is it the same either way?

   _____

   e. In which case is the stopping force greater? Or is it the same either way?

   _____

4. Use your findings from this activity to explain the purpose of airbags in cars. Don't use words like "cushion," or "absorb." Do use terms like "stopping time," and "stopping force."

_____

_____

5. What are some other examples of changing stopping time to change stopping force?

_____

_____

Laboratory Manual for *Conceptual Integrated Science,* © 2007 Addison Wesley

CONCEPTUAL INTEGRATED SCIENCE	Activity

Momentum and Energy: Impulse–Momentum Relationship

Bouncy Board

Purpose
In this activity, you will investigate the effect of stopping time when momentum changes.

Required Equipment and Supplies
table
meterstick
short length [30 to 40 centimeters (cm)] of weak string (e.g.: mailing parcel twine)
various masses

Discussion
In bungee jumping, it is important that the cord supporting the jumper stretches. If the cord has no stretch, then when fully extended it either brings the jumper to a sudden halt or the cord snaps. In either case, ouch! Whenever a falling object is brought to a halt, the force that slows the fall depends on the time it acts. We will see evidence of this in this activity.

Procedure
Through the small hole at the end of a meterstick, thread a piece of string about 30 cm long and tie it in place. (If there is no hole, consider drilling one, or tying the string very tightly. Attach a mass to the other end of the string—try a kilogram for starters. Place the stick on a tabletop and slide the stick so that most of its length extends over the edge of the table (Figure 1). While holding the stick firm to the table, hold the mass slightly beyond the edge of the stick—then drop the mass. The string and the bending of the stick should stop its fall. The string shouldn't break (if it does break, try a stronger string or a smaller mass—experiment). Now repeat this activity, but with only a small portion of the stick extending over the table's edge (Figure 2). What happens now?

Try different configurations to see what conditions result in string breaking. Experiment with different masses, different amounts of meterstick overhang, and different string lengths.

Figure 1

Figure 2

Summing Up

1. Why is it important that a bungee jumper be brought to a halt gradually?

_____

_____

2. How does the *impulse = change in momentum* formula, *Ft = mv*, apply to this activity?

_____

_____

3. Exactly why did the string break when there was less "give" to the meterstick?

_____

_____

4. The breaking strength of the string most certainly plays a role in this activity. How does the length of the string play a role?

_____

_____

5. How does the falling mass play a role? Will twice the mass require twice the stopping force if it is brought to a halt in the same time? Defend your answer.

_____

_____

6. Why is it important that fishing rods bend?

_____

_____

Laboratory Manual for *Conceptual Integrated Science*, © 2007 Addison Wesley

Name _____ Date _____

| **CONCEPTUAL INTEGRATED SCIENCE** | **Experiment** |

Momentum and Energy: The Work-Energy Theorem
An Uphill Climb

Purpose
In this experiment, you will determine what advantage—if any—there is in using an inclined plane to move an object to a higher elevation.

Required Equipment and Supplies
dynamics cart and track or board that can be inclined and secured at various angles
spring scale (capable of weighing the dynamics cart)
table clamp
support rod
rod clamp
meterstick
protractor

Discussion
Why are ramps used when lifting heavy objects? Does it make the task easier (requiring less force)? Does it make the movement shorter (requiring less distance)? Does it make the effort more efficient (requiring less work)? Perhaps it does several of these; maybe it does none of them. You will learn more from this lab if you record your initial thoughts before making any measurements or calculations.

What advantages or disadvantages are there in using a ramp when lifting a heavy object?

_____

_____

_____

In this experiment, the cart will act as the heavy object. Your task will be to move your cart a vertical distance of 20 cm above the tabletop. You will arrange a series of ramps (inclined planes) at different angles to accomplish this task. You will measure the force needed to move a cart up the incline. You will also measure the distance through which that force would be applied to finish the job. You will then calculate the work required to lift an object using an inclined plane. By the end of the experiment, you will be able to identify what an inclined plane can do for you in terms of force, distance, and work.

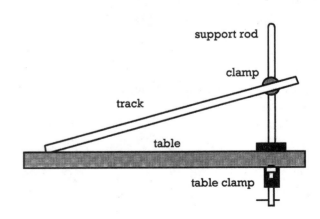

support rod

clamp

track

table

table clamp

Procedure

Shallow Incline
Step 1: Arrange the apparatus as shown in Figure 1. The plane should be inclined at an angle between 20° and 30°. Check the angle with the protractor as shown in Figure 2.

Figure 1

Step 2: Measure the force needed to pull the cart along this incline using the spring scale as shown in Figure 3. The spring scale is held parallel to the inclined plane when the measurement is made. Because the force needed to move the cart at a constant speed is the same as the force needed to keep the cart at rest on the plane, measure the force when the cart is at rest. Record the force below and transfer the value to the data table.

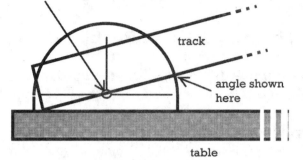

Figure 2

Shallow incline force:

F = _____ N

Step 3: Measure the distance the cart would travel along the inclined path to get from the tabletop to a distance 20 cm above the tabletop. See Figure 4. The path starts at the tabletop (even if the object being used as an incline plane doesn't come all the way down to the tabletop). The path ends where the inclined plane is 20 cm above the tabletop. This distance will be greater than 20 cm for all inclined planes (unless the plane goes straight up). Convert the distance from centimeters to meters. Record the distance below and transfer the value (in meters) to the data table.

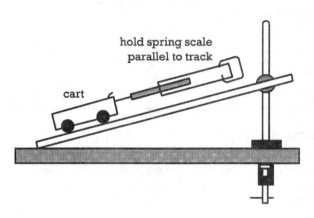

Figure 3

Shallow incline distance:

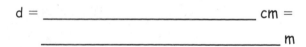

d = _____ cm =

_____ m

Medium Incline
Step 4: Increase the angle of incline to a value between 40° and 50°.

Step 5: Measure the force needed to move the cart along this incline and record it on the data table.

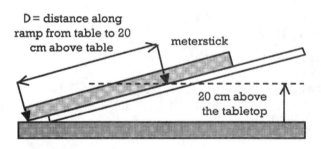

Figure 4

Step 6: Measure the distance the cart would travel along this path to get 20 cm above the tabletop. The upper end of the incline is much higher now than it was for the shallow incline, but our only concern is get the cart 20 cm above the tabletop. The distance the cart would travel along this path will be smaller than the distance it would travel along the shallow incline. Record the distance (in meters) on the data table.

Steep Incline
Step 7: Increase the angle of incline to a value between 60° and 70°.

Step 8: Measure the force needed to move the cart along this incline and record it on the data table.

Step 9: Measure the distance the cart would travel along this path to get 20 cm above the tabletop. Record the distance (in meters) on the data table.

Straight Up
Step 10: Measure the force needed to move the cart straight up. No incline is needed for this task. Record the force on the data table.

Step 11: The distance the cart would travel along this path to get 20 cm above the tabletop is 20 cm. Record the distance (in meters) on the data table.

Calculating Work
Step 12: Calculate the work done along the shallow incline. Multiply the force applied to the cart by the distance the cart traveled to get 20 cm above the tabletop. Show the calculation for the shallow path below and transfer the result to your data table. Include the correct units in all your values.

$W = F \cdot d =$ _____ $\times$ _____ $=$ _____

Step 13: Calculate the work for the other paths: the medium incline, steep incline, and straight up. Record the results on the data table.

Step 14: Show your instructor your completed data table before proceeding to the Summing Up section.

Data Table

	Angle θ (degrees)	Force F (newtons)	Distance d (meters)	Work W (joules)
Shallow Path				
Medium Path				
Steep Path				
Straight Up				

Summing Up

1. As the incline gets steeper, what happens to the force required to pull the cart?

   _____ The force increases significantly.

   _____ The force decreases significantly.

   _____ The force remains about the same.

   (Compare the force needed to pull the cart along the shallow path to the force needed to pull the cart straight up. A significant difference is one that is 20% or greater.)

2. As the incline gets steeper, what happens to the distance traveled by the cart?

   _____ The distance increases significantly.

   _____ The distance decreases significantly.

   _____ The distance remains about the same.

   (Compare the distance along the shallow path to the distance of the path straight up.)

3. As the incline gets steeper, what happens to the work required to move the cart 20 cm above the tabletop?

   _____ The work increases significantly.

   _____ The work decreases significantly.

   _____ The work remains about the same.

   (Compare the work needed to move the cart up the shallow path to the work needed to move the cart straight up.)

4. What is the *advantage* of using an inclined plane rather than moving something straight up?

   _____

   _____

   _____

5. What is the *disadvantage* if using an inclined plane rather than moving something straight up?

   _____

   _____

   _____

6. The work done to move something is a measure of the energy required to complete the task. The energy required to move an automobile is provided by the fuel. Would it be more fuel efficient to drive to the top of a hill along a steeply inclined road or a gradually inclined road? Explain your answer in terms of what you observed in this experiment.

   _____

   _____

   _____

Name _____ Date _____

| **CONCEPTUAL INTEGRATED SCIENCE** | **Experiment** |

Momentum and Energy: Conservation of Energy

Rolling Stop

Purpose
In this experiment, you will investigate the relationship between the height of a ball rolling down an incline and its stopping distance when it rolls off the incline.

Required Equipment and Supplies
about a 2-meter (m) ramp (e.g.: 5/8 inch aluminum channel) with support about 1/2 m high
steel ball
meterstick
carpet floor, or piece of **flat** carpet strip [about 4 m long and about 50 centimeters (cm) wide]

Discussion
Mechanical energy is the product of force and distance. When we exert a force to change the energy of an object, we do *work* on an object. A measure of that work, or change in energy, is force times distance. Elevate a ball and we do work on it. With a force equal to its weight, we lift it a certain height against gravity. The work done gives it *energy of position*—gravitational potential energy. Raise it twice as high and it has twice the energy. Another way of saying it has twice the energy is to say it has twice the ability to *do* work. When it rolls to the bottom of the incline, it can do work on whatever it interacts with. If it has twice the energy, it can do twice the work. An easier-to-visualize example is that of a crate sliding onto a factory floor. In sliding, it does work on the floor and heats it up as it skids. This work is the force of friction ¥ distance of sliding. The question is raised: Will it skid twice as far if it has twice the energy? We'll answer this question not by sliding crates down a ramp (for much of their potential energy would go into heating the ramp, which complicates matters), but by rolling balls down inclines and then onto a carpet.

Procedure
Step 1: Mark the ramp at 30 cm, 60 cm, 90 cm, and 120 cm from the bottom end. Assemble the ramp so that when you roll a ball down it, the length of carpet it rolls onto is sufficient to stop it. Experiment to see how far this is (Figure 1).

Figure 1

Step 2: Release the steel ball at each of the intervals along the ramp. Measure the vertical height from the floor or table. Roll the ball three times from each height and record the stopping distances in the Step 2 Data Table.

Step 3: Change the angle of the ramp, but launch the ball from the **same vertical height** that you did for the previous ramp position (Figure 2).

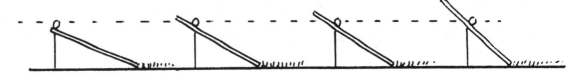

Figure 2

Step 2 Data

Initial Position of Ball (cm)	Initial Height of Ball (cm)	Stopping Distance (cm)			Average Stopping Distance (cm)
		Trial 1	Trial 2	Trial 3	
30					
60					
90					
120					

Step 3 Data

Initial Position of Ball (cm)*	Initial Height of Ball (cm)**	Stopping Distance (cm)			Average Stopping Distance (cm)
		Trial 1	Trial 2	Trial 3	
30					
60					
90					
120					

*Because the angle is different, these will no longer be 30 cm, 60 cm, 90 cm, and 120 cm.

**These values must be the same as the values in the Step 2 Data table.

Going Further

On the graph paper provided by your instructor, construct a graph of average stopping distance (vertical axis) versus height (horizontal axis) for the ball in Step 2.

Summing Up

1. Do the graphs indicate direct proportions between height of release and stopping distance?

_____

2. Would you get the same relationship between distance up the ramp (parallel to the ramp rather than height) and stopping distance? \ or why not?

_____

3. How did the stopping distances for the different balls compare? How did the relationship between release heights and stopping distances compare?

_____

4. Nowhere in this activity is there a mention of kinetic energy—energy of motion (*KE*). A little study of energy conservation tells us that the gain in *KE* of the ball as it rolls down the ramp is equal to the decrease in *PE* as the ball loses height—in short, *PE* = *KE*. Why were we able to bypass *KE* in our analysis here? (The answer to this question underlies the reason that physics types use energy principles to solve problems—intermediate steps can be skipped!)

_____

_____

 Laboratory Manual for *Conceptual Integrated Science,* © 2007 Addison Wesley

| CONCEPTUAL INTEGRATED SCIENCE | Experiment |

Momentum and Energy: Conservation of Energy

Dropping the Ball

Purpose
In this experiment, you will lift a ball and drop it. You will determine and compare the potential energy of the ball before it's dropped to the kinetic energy of the ball right before it hits the ground.

Required Equipment and Supplies
acrylic tube (1.0-inch diameter, about 4 feet in length)
steel ball, about 16 millimeters (mm)
small rare-earth magnet (neodymium or equivalent)
2 photogates and timers
table clamp
support rod
2 three-finger clamps or buret clamps
meterstick

SAFETY NOTE: Use caution when handling the magnet to avoid pinching. Keep it away from computers, sensitive electronic devices, and magnetic storage media such as computer disks.

Discussion
When an object is lifted, the work done to lift the object is transformed into potential energy. If the object is then dropped, the potential energy is transformed to kinetic energy as it falls. In this experiment, we will determine the potential energy of a lifted object. We will also determine the kinetic energy of a dropped object. We will then compare the potential energy that an object has when it is lifted to a certain height to the kinetic energy it has after it has fallen from that height.

Procedure
Step 1: Arrange the apparatus as shown in Figures 1 and 2. The upper photogate (gate 1) should be about 5 centimeters (cm) above the lower photogate (gate 2). Connect both gates to the timer.

Step 2: Configure the photogate timer to read the time between the two gates (that is, the timer starts when the beam of the first photogate is interrupted and stops when the beam of the second photogate is interrupted).

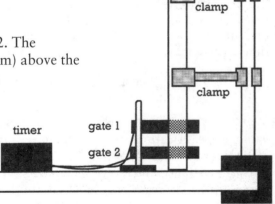

Figure 1. Side View

Figure 2. Top View
Make sure the photogate beam passes through the diameter of the tube.

Step 3: Measure the distance between the photogate beams as shown in Figure 3. **Be very careful in this measurement.** Record the distance in centimeters and convert it to meters.

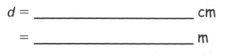

Distance between photogate beams:

$d =$ _____ cm

$=$ _____ m

Step 4: Determine the mass of the ball. Record its mass in grams and kilograms.

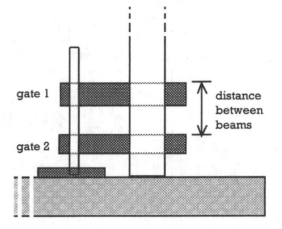

Figure 3

Mass of steel ball:

$m =$ _____ g

$=$ _____ kg

Step 5: Set the ball inside the tube. Use the magnet to lift the ball up through the tube so that the bottom of the ball is 40 cm above the upper photogate as shown in Figure 4. Be careful: the bottom of the ball must be 40 cm above the **upper photogate,** not 40 cm above the **table.**

Step 6: Clear or "arm" the photogate timer so that it is prepared to make a measurement.

Step 7: Carefully remove the magnet from the side of the tube. Doing so will release the ball to fall to the bottom of the tube. When the ball passes through the photogate beams, a measurement will be made and displayed on the timer. The goal is to release the ball from rest, so take care not to give the ball upward or downward motion when you release it.

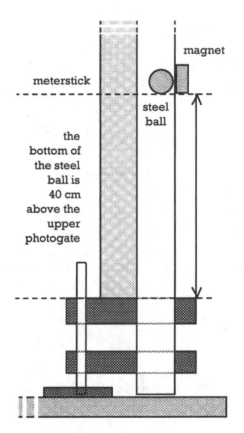

Figure 4

Step 8: If the trial went well, record the time value. Repeat the process until you have three reliable time values. (If you make a mistake during a trial, do not record the result. Simply repeat the process until you have a good trial.)

Time values: _____ _____ _____

Step 9: Determine the average of the three values and record it on the data table.

Step 10: Repeat the process for drops from 60 cm, 80 cm, and 100 cm. Record the results on the data table.

Laboratory Manual for *Conceptual Integrated Science*, © 2007 Addison Wesley

Data and Calculations

Drop Height h [meters (m)]	Photogate Time t [seconds (s)]	Speed v [meters/second (m/s)]	Potential Energy PE [joules (J)]	Kinetic Energy KE (J)
0	—	0	0	0
0.40				
0.60				
0.80				
1.00				

1. Calculate the potential energy of the ball when it was 40 cm above the photgate using the equation $PE = mgh$ (m is the mass of the ball, g is 9.8 m/s^2, and h is 0.40 m). Show your work and your solution. (The value should be between 0.050 J and 0.100 J.) Record your solution in the data table as well.

_____

_____

_____

2. Calculate the potential energy of the ball at the other heights and record your solutions in the data table.

3. Calculate the speed of the ball after it fell 40 cm using the equation $v = d/t$ (d is the distance between the photogates, and t is the time it took the ball to pass between the photogates). Show your work and your solution. (The value should be between 2.50 m/s and 3.00 m/s.) Record your solution in the data table.

_____

_____

_____

4. Calculate the speed of the ball after falling from the other drop heights and record your solutions on the data table.

5. Calculate the kinetic energy of the ball after it fell 40 cm using the equation $KE = 1/2mv^2$ (m is the mass of the ball, and v is the speed of the ball). Show your work and your solution. (The value should be between 0.050 and 0.100 J.) Record your solution in the data table.

_____

_____

_____

6. Calculate the kinetic energy of the ball after falling from the other drop heights and record your solution on the data table.

Summing Up

1. When the drop height doubles from 40 cm to 80 cm, which of the following quantities also doubles (approximately)?

_____ speed after the fall

_____ potential energy at the drop height

_____ kinetic energy after the fall

2. Which statement best describes the relationship between the potential energy at the drop height to the kinetic energy after the fall?

_____ The potential energy is always significantly higher than the kinetic energy.

_____ The kinetic energy is always significantly higher than the potential energy.

_____ The potential energy and kinetic energy are about the same.

(A significant difference in this experiment would be a difference of 20% or more.)

3. Use your findings to predict the following values for a trial involving dropping the ball from 160 cm.

a. Potential energy = _____

b. Kinetic energy = _____

c. Speed after falling = _____

(Hint: By what factor was the "80-cm ball" faster than the "40-cm ball"? Use this factor to determine how much faster the "160-cm ball" will be compared to the "80-cm ball.")

Going Further

When the ball was released from a particular height, its potential energy was transformed to kinetic energy as it fell. This type of energy transformation happens on a roller coaster as well. Potential energy that the roller coaster has at the top of the first hill is transformed to kinetic energy as it rolls downward. In an ideal roller coaster (with no frictional losses), the kinetic energy at the bottom of the hill would be equal to the potential energy at the top.

Consider an ideal roller coaster. The first hill has a certain height. When the roller coaster reaches the bottom of the hill, it is traveling at a certain speed.

1. How much higher would the hill have to be so that the roller coaster had twice as much **kinetic energy** at the bottom of the hill?

_____

2. How much higher would the hill have to be so that the roller coaster had twice as much **speed** at the bottom of the hill? (The answer to this question is not the same as the answer to the previous question.)

_____

3. In real-world roller coasters, each hill is shorter than the hill before it. Why do you suppose that is?

_____

_____

Laboratory Manual for *Conceptual Integrated Science*, © 2007 Addison Wesley

Name _____ Date _____

| CONCEPTUAL INTEGRATED SCIENCE | Experiment |

Gravity: Weight and Weightlessness

The Weight

Purpose
In this activity, you will investigate the relationship between weight and mass.

Required Equipment and Supplies
spring scale (10-newton capacity)
slotted masses [one 100 gram (g), two 200 g, one 500 g]
mass hanger
table clamp
support rod
rod clamp
crossbar (short rod)
collar hook
graph paper

Discussion
Mass and weight are different quantities. Mass is a measure of an object's inertia, the extent to which an object resists changes to its state of motion. Weight is a measure of the interaction between an object and the planet the object is nearest to. Usually that planet is the earth. The weight of an object is related to its mass. In this activity, we will find out what that relationship is.

Procedure
Step 1: Check the calibration of the spring scale and adjust it if necessary. The spring scale needs to read zero when there is no load on it. Ask your instructor for instructions on how to calibrate the spring scale.

Step 2: Arrange the apparatus as shown in Figure 1.

Step 3: Determine the mass (in grams) and the weight (in newtons—as shown on the spring scale) of the mass hanger. Record those values on the second row of the data table.

Step 4: Add 100 g of slotted mass to the mass hanger. The total mass of the load on the spring scale is now 100 g plus the mass of the mass hanger. Record the total mass and the weight value (shown on the spring scale) on the next row of the data table.

Step 5: Repeat the previous step with 200 g, 300 g, 400 g, 500 g, 600 g, 700 g, and 800 g of slotted mass. Remember that the total mass in each case is the sum of the slotted mass and the mass of the mass hanger. When you are done, the column of mass values (in grams) and the corresponding column of weight values (in newtons) will be filled.

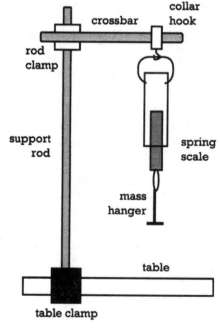

Figure 1

Data

Total Hanging Mass m (grams)	Total Hanging Mass m (kilograms)	Weight of Mass W (newtons)
0	0	0

Step 6: Convert the mass values to kilograms (1000 g = 1 kg, so 237 g = 0.237 kg, etc).

Step 7: Make a plot of weight vs. mass, using the mass values in kilograms. As is always the case, the first variable listed in the title of the graph constitutes the vertical axis and the second variable constitutes the horizontal axis. Generally speaking, the first variable listed is the dependent variable (the one you measure during the activity), and the second variable is the independent variable (the one you control during the activity).

Step 8: Title the graph, label the axes to correctly indicate the quantity, units, and scale of each axis.

Step 9: Draw a straight line of best fit through the data points plotted on your graph.

1. Determine the slope of your best-fit line and record it below.

   Slope: _____

2. What are the units of the slope you found? The slope does have units! Show the units in your answer to the previous question.

The slope of the graph is the relationship between weight and mass. The slope tells how many newtons of weight pull down on each kilogram of mass.

 Laboratory Manual for *Conceptual Integrated Science,* © 2007 Addison Wesley

Summing Up

1. What would have been the weight of 1.0 kg? Extend your best-fit line or use a ratio to determine the answer.

_____

_____

2. On the moon, each kilogram of mass is pulled down with 1.6 newtons of weight. Add a dashed line to your graph showing the results if this activity had been done on the moon. How does the slope of the moon line compare to the slope of the earth line? Is it steeper (more vertical) or shallower (more horizontal)?

_____

_____

_____

3. If the activity had been done on Jupiter, the resulting line would have had a steeper (more vertical) slope. What does this tell you about the strength of Jupiter's gravitational field as compared to Earth's?

_____

_____

_____

Name _____ Date _____

CONCEPTUAL INTEGRATED SCIENCE	Activity

Gravity: The Legend of the Falling Apple

Reaction Time

Purpose

In this activity, you will measure your personal reaction time.

Required Equipment and Supplies

dollar bill
centimeter ruler

Discussion

Reaction time is the time interval between receiving a signal and acting on it—for example, the time between a tap on the knee and the resulting jerk of the leg. Reaction time often affects the making of measurements. Consider using a stopwatch to measure the time for a 100-meter dash. The watch is started after the gun sounds and is stopped after the tape is broken. Both actions involve reaction time.

Procedure

Step 1: Hold a dollar bill so that the midpoint hangs between your partner's fingers. Challenge your partner to catch it by snapping his or her fingers shut when you release it.

The distance the bill will fall is found using

$$d = \frac{1}{2} at^2$$

Simple rearrangement gives the time of fall in seconds.

$$t^2 = \frac{2d}{g}$$

$$t = \sqrt{\frac{2}{980}} \sqrt{d}$$

$$t = 0.045 \sqrt{d}$$

[For d in centimeters (cm) and t in seconds (s), we use $g = 980$ cm/s^2.]

Step 2: Have your partner similarly drop a centimeter ruler between your fingers. Catch it and note the number of centimeters that passed during your reaction time. Then calculate your reaction time using the formula

$$t = 0.045 \sqrt{d}$$

where d is the distance in centimeters.

Reaction time = _____

Summing Up

1. What is your evidence for believing or disbelieving that your reaction time is always the same? Is your reaction time different for different stimuli?

_____

_____

2. Suggest some possible explanations for why reaction times are different for different people.

_____

_____

3. When might reaction time significantly affect measurements you might make using instruments for this course? How could you minimize its role?

_____

_____

4. What role does reaction time play in applying the brakes to a car in an emergency situation? Estimate the distance a car travels at 100 km/h during your reaction time in braking.

_____

_____

5. Give examples of how reaction time is important in sports.

_____

_____

CONCEPTUAL INTEGRATED SCIENCE	Activity

Gravity: Projectile Motion

The Big BB Race

Purpose
In this activity, you will compare the path of a projectile launched horizontally with that of an object in free fall.

Required Equipment and Supplies
simultaneous launcher/dropper

Discussion
Suppose a ball bearing (BB) were launched horizontally at the same time another BB were dropped from the same height. Which one would reach the ground first?

Procedure
Step 1: In Figure 1 draw the path you think the launched projectile will take. That is, draw a line connecting the launch point and the impact point in the diagram that traces the path you think the BB will follow.

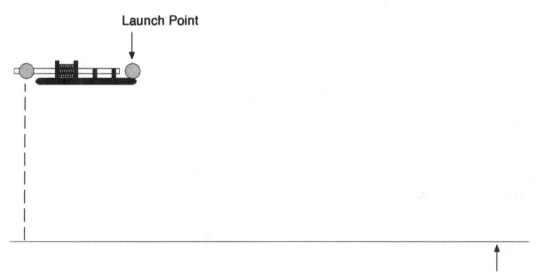

Figure 1

Step 2: Consider the predictions of three students. Write arguments supporting each prediction (whether you agree with the prediction or not).

Student X predicts the dropped BB will hit first. Why might Student X believe this?

_____

Student Y predicts the fired BB will hit first. Why might Student Y believe this?

Student Z predicts both balls will hit at the same time. Why might Student Z believe this?

Step 3: One of the reasons sometimes offered to support the prediction that the dropped ball will hit first is that the launched ball will travel forward for some distance before starting to move downward. What factors might determine the length of this "no-fall distance," as shown in Figure 2?

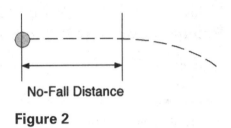

No-Fall Distance

Figure 2

Step 4: Which prediction do you agree with: dropped ball hits first, fired ball hits first, or both hit at the same time?

Step 5: Observe the operation of the simultaneous launch/drop mechanism.
   a. Observe a dropped BB.
   b. Observe a launched BB.
   c. Observe the Big BB Race—a simultaneous launch *and* drop.

Step 6: Which BB hit the ground first, or was it a tie?

Summing Up
1. How does the horizontal motion of a projectile affect the vertical motion of the projectile? In other words, does the horizontal motion of the projectile make it move faster or slower in the vertical direction (or does it have no effect)?

2. Which factors—if any—appear to have the greatest effect on the no-fall distance discussed above?

3. If the launched BB had a rocket engine propelling it forward after it was launched, what—if anything—would have been different about the outcome of the Big BB Race?

Laboratory Manual for *Conceptual Integrated Science,* © 2007 Addison Wesley

CONCEPTUAL INTEGRATED SCIENCE | Activity

Gravity: Projectile Altitude and Range

Bull's Eye

Purpose

In this experiment, you will predict the landing point of a projectile.

Required Equipment and Supplies

1/2" (or larger) steel ball
empty can
meter stick
stopwatch
means of projecting the steel ball horizontally at a known velocity

Discussion

The concept is that a projectile moves with constant speed in the horizontal direction while undergoing free fall acceleration in the vertical direction, as the sketch of the ball toss in Figure 1 shows. The horizontal and vertical motions are **independent** of each other!

Instead of tossing a ball from a cliff, you'll fire a steel ball off the edge of a table and into a can on the floor below—all without a trial shot!

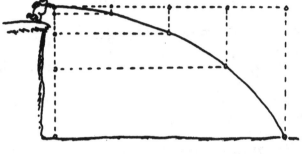

Figure 1

When engineers build bridges or skyscrapers, they do *not* do so by trial and error. For the sake of safety and economy, the effort must be right the *first* time. Your goal in this experiment is to predict where a steel ball will land when projected horizontally from the edge of a table. The final test of your measurements and computations will be to position an empty can so that the ball lands in the can on the *first* attempt.

Procedure

Step 1: Your instructor will provide you a means of projecting the steel ball at a known horizontal velocity (it may be a spring gun, a ramp, or some other device). Position the ball projector at the edge of a table so the ball will land downrange on the floor. Do **not** make any practice shots (the fun of this experiment is to *predict* where the ball will land without seeing a trial). Figure 2 shows the quantities involved in this experiment.

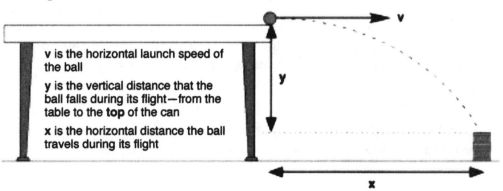

v is the horizontal launch speed of the ball

y is the vertical distance that the ball falls during its flight—from the table to the **top** of the can

x is the horizontal distance the ball travels during its flight

Figure 2

You will be given the initial speed of the steel ball (or perhaps you'll have to devise a way to measure it). Record the firing speed here.

Horizontal speed v = _____ cm/s

Step 2: Carefully measure the vertical distance y the ball must drop from the bottom end of the ball projector in order to land in an empty soup can on the floor. Be sure to take the height of the can into account when you make this measurement.

Vertical distance y = _____ cm

Step 3: The vertical free fall motion is governed by the equation $y = (1/2)gt^2$. Use the rearranged form of that equation, $t = \sqrt{(2y/g)}$, calculate the time t it takes the ball to fall from its initial position to the can. For gravitational acceleration, use $g = 980$ cm/s^2.

Time of flight t = _____ s

Step 4: The range is the horizontal distance a projectile travels, x. Predict the range of the ball using $x = vt$. Write down your predicted range.

Predicted range x = _____ cm

Now place the can on the floor where you predict it will catch the ball.

Step 5: Only after your instructor has checked your predicted range and your can placement, shoot the ball.

Summing Up

1. Did the ball land in the can on the first trial? If not, how many trials were required?

_____

2. What possible errors would account for the ball overshooting the target?

_____

_____

3. What possible errors would account for the ball undershooting the target?

_____

_____

4. If the can you used were replaced with a taller can, would you get a successful run if the ball started with a higher speed, lower speed, or the same speed? Explain.

_____

_____

CONCEPTUAL INTEGRATED SCIENCE	Activity

Heat: The Kinetic Molecular Theory of Matter

Dance of the Molecules

Purpose
In this activity, you will investigate the difference between hot water and cold water—on the **molecular** level.

Required Equipment and Supplies
source of cold water
source of hot water
two empty baby food jars or small beakers
food coloring

Discussion
The difference between hot things and cold things is referred to as *temperature*. The temperature of an object depends on the kinetic energy of the random motions of its molecules. A thermometer can be used to measure temperature, but we can't always tell a hot thing from a cold thing just by looking at it. Consider a glass of hot water and a glass of cold water side by side. The water molecules in each glass don't appear to be moving differently from each other. But they are. Follow the steps below to **see** the difference.

Procedure
Step 1: Fill one jar with cold water and the other with hot water.

Step 2: Spend a minute or two making a prediction. What will happen if a few drops of food coloring are added to each jar of water? Record your prediction using words and pictures. (Let the water stand while you record your prediction. This will allow swirling currents and turbulence in the water to diminish.)

_____

_____

_____

Step 3: Carefully add just a drop or two of food coloring to each jar. Watch the coloring spread out for 1 minute before recording your observations, taking care not to disturb the water in any way. Record your observations using words and pictures.

_____

_____

_____

Summing Up

1. How do your observations support the principle that the molecules in hot objects move faster than the molecules in cold objects?

_____

_____

2. Suppose you let the experiment continue for several hours. What would the jars of water and food coloring look like afterward?

_____

_____

3. Consider a fragrant candle burning in the corner of one side of a large room. It would be possible to smell the fragrance on the other side of the room after a short period of time. Use your observations to explain how this is so.

_____

_____

4. Would the fragrance get to you more quickly, more slowly, or in the same amount of time if the air in the room were colder?

_____

_____

 Laboratory Manual for *Conceptual Integrated Science,* © 2007 Addison Wesley

CONCEPTUAL INTEGRATED SCIENCE

Experiment

Heat: Specific Heat Capacity

Temperature Mix

Purpose

In this activity, you will predict the final temperature of a mixture of cups of water at different temperatures.

Required Equipment and Supplies

3 Styrofoam cups
liter container
thermometer (Celsius)
pail of cold water
pail of hot water

Discussion

If you mix a pail of cold water with a pail of hot water, the final temperature of the mixture will be between the two initial temperatures. What information would you need to predict the final temperatures? You'll begin with the simplest case of mixing *equal* masses of hot and cold water.

Procedure

Step 1: Begin by marking your three Styrofoam cups equally at about the three-quarter mark. You can do this by pouring water from one container to the next and marking the levels along the inside of each cup.

Step 2: Fill the first cup to the mark with hot water from the pail, and fill the second cup with cold water to the same level. Measure and record the temperatures of both cups of water.

Temperature of cold water = _____

Temperature of hot water = _____

Step 3: Predict the temperature of the water when the two cups are combined. Then pour the two cups of water into the liter container, stir the mixture slightly, and record its temperature.

Predicted temperature = _____

Actual temperature of water = _____

1. If there was a difference between your prediction and your observation, what may have caused it?

_____

Pour the mixture into the sink or waste pail. (Don't be a klutz and pour it back into either of the pails of cold or hot water!) Now we'll investigate what happens when *unequal* amounts of hot and cold water are combined.

Step 4: Fill one cup to its mark with cold water from the pail. Fill the other two cups to their marks with hot water from the other pail. Measure and record their temperatures. Predict the temperature of the water when the three cups are combined. Then pour the three cups of water into the liter container, stir the mixture slightly, and record its temperature.

Predicted temperature = _____

Actual temperature of water = _____

Pour the mixture into the sink or waste pail. Again, do *not* pour it back into either of the pails of cold or hot water!

2. How did your observation compare with your prediction?

_____

3. Which of the water samples (cold or hot) changed more when it became part of the mixture? In terms of energy conservation, suggest a reason for why this happened.

_____

Step 5: Fill two cups to their marks with cold water from the pail. Fill the third cup to its marks with hot water from the other pail. Measure and record their temperatures. Predict the temperature of the water when the three cups are combined. Then pour the three cups of water into the liter container, stir the mixture slightly, and record the temperature.

Predicted temperature = _____

Actual temperature of water = _____

Pour the mixture into the sink or waste pail. (By now you and your lab partners won't alter the source temperatures by pouring waste water back into either of the pails of cold or hot water.)

4. How did your observation compare with your prediction?

_____

5. Which of the water samples (cold or hot) changed more when it became part of the mixture? Suggest a reason for why this happened.

_____

_____

Summing Up

6. What determines whether the equilibrium temperature of a mixture of two amounts of water will be closer to the initially cooler or warmer water?

_____

7. How does the formula $Q = mc\Delta T$ apply here?

_____

_____

Name _____ Date _____

Heat: Specific Heat Capacity

Spiked Water

Purpose
In this experiment, you will determine which is better able to increase the temperature of a quantity of water: a mass of hot nails or the same mass of equally hot water.

Required Equipment and Supplies
equal arm balance (Harvard trip balance or equivalent)
4 large insulated cups
bundle of short, stubby nails tied together with string
thermometer (Celsius)
hot and cold water
paper towels

Discussion
Suppose you have cold feet when you go to bed, and you want something to keep your feet warm throughout the night. Would you prefer to have a bottle filled with hot water, or one filled with an equal mass of nails at the same temperature as the water? The one that can store more thermal energy will do the better job. But which one is it? In this experiment, we'll find out.

Procedure
1. Consider the equal masses of nails and water, both warmed to the same temperature. Which one will do a better job of heating a sample of cold water? Make a prediction before carrying out the experiment.

_____

_____

Step 1. Fill two cups 1/3 full of **cold** water. Place one cup on each pan of the balance to make sure they contain equal masses of water, as shown in Figure 1. Add water to the lighter side until they do. Make sure your bundle of nails can be completely submerged in the cold water.

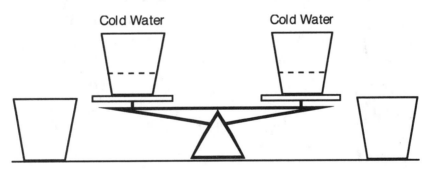

Cold Water   Cold Water

Figure 1

Step 2: Dry the nails.

Step 3: Place a large empty cup on each pan of a beam balance. Place the bundle of nails into one of the cups. Add **hot** water to the other cup until it balances the cup of nails, as shown in Figure 2. When the two cups are balanced, the mass in each cup is the same.

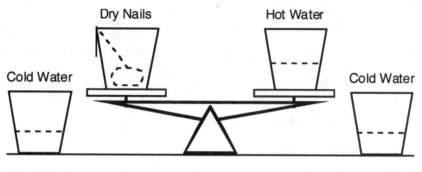

Figure 2

Step 4: Lower the bundle of nails into the hot water as shown in Figure 3. Be sure that the nails are completely submerged by the hot water. Allow the nails and the water to reach thermal equilibrium. (This will take a minute or two.)

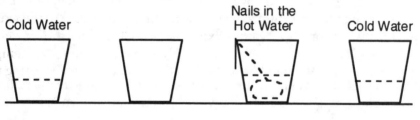

Figure 3

Step 5: Once the nails and hot water have come to thermal equilibrium, take the nails out of the hot water and put them in one of the cups of cold water. Pour the remaining hot water into the other cup of cold water. See Figure 4.

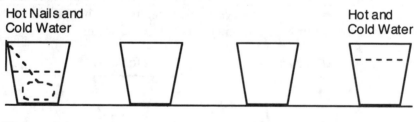

Figure 4

Step 6: Measure and record the final temperature of the mixed water.

$T_W = $ _____ °C

Step 7: When the temperature of the nail–water mix stops rising, measure and record it.

$T_N = $ _____ °C

Summing Up

1. Which was hotter before being put into the cold water, the nails or the hot water the nails were soaking in? How do you know?

_____

_____

2. Which was more effective in raising the temperature of the cold water, the hot nails or the hot water?

_____

_____

3. Suppose you have cold feet when you go to bed, and you want something to warm your feet throughout the night. Would you prefer to have a bottle filled with hot water, or one filled with an equal mass of nails at the same temperature as the water? Relate your answer to your findings in this experiment.

_____

_____

4. A student who conducted this experiment suggested that the temperature of the nail–water mix rose more than it should have because some water clung to the nails when they were transferred from the hot water to the cold water. Another student says that this caused the temperature of the nail–water mix to rise less than it should have if no water clung to the nails during the transfer. Whom do you believe and why?

_____

_____

_____

_____

CONCEPTUAL INTEGRATED SCIENCE	Experiment

Heat: Specific Heat Capacity

Specific Heat Capacities

Purpose
In this activity, you will measure the specific heat capacities of some common metals.

Required Equipment and Supplies
hot plate
metal specimens
beaker
tongs
Styrofoam cups
balance
thermometer

Discussion
Have you ever held a hot piece of pizza by its crust only to have the moister parts burn your mouth when you take a bite? The meats and cheese have high specific heat capacity, whereas the crust has a low specific heat capacity. How can you compare the specific heat capacities of different materials?

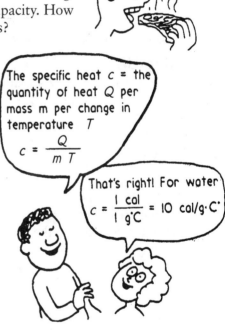

In this experiment, you will increase the temperature of metal specimens to that of boiling water. Then you'll place each specimen in a double Styrofoam cup that contains a mass of room-temperature water equal to the mass of each specimen. The specimen will cool and the water temperature will rise until both are at the same temperature. That is, the heat lost by the specimen equals the heat gained by the water.

The specific heat c = the quantity of heat Q per mass m per change in temperature T

$$c = \frac{Q}{m\,T}$$

That's right! For water
$$c = \frac{1 \text{ cal}}{1 \text{ g°C}} = 10 \text{ cal/g·C°}$$

$$Q_{lost} = Q_{gained}$$

$$c_s m_s \Delta T_s = c_w m_w \Delta T_w$$

The specific heat capacity of the specimen is

$$c_s = \frac{c_w m_w \Delta T_w}{m_s \Delta T_s}$$

For water $c_w = 1.00$ cal/g · °C and if the mass of water is the same as that of the mass of the specimen, then the specific heat of the sample is simply the ratio of the temperatures:

$$c_s = 1.00 \frac{\text{cal}}{\text{g·°C}} \left(\frac{\Delta T_w}{\Delta T_s} \right)$$

Procedure
Step 1: Measure the mass of each specimen, and record their masses in Table 1. Then place the specimens in a beaker of water, deep enough to cover the specimens, and heat the water to boiling.

Step 2: While the water in the beaker is being brought to a boil, assemble as many pairs of Styrofoam cups with one cup inside the other as you have specimens. You have just constructed inexpensive double-walled calorie meters, called *calorimeters*. Because 1 milliliter (mL) of water has a mass of 1 gram (g), carefully measure as many milliliters of tap water as there are grams for each specimen and place the measured water in each calorimeter. Measure the temperature of this water in the calorimeters with a thermometer.

Step 3: Let the water in the beaker boil vigorously for more than a minute until you are convinced the specimens are in thermal equilibrium with the boiling water. Then using tongs, quickly remove each specimen from the boiling water, shake away excess droplets of water, and place each in the appropriate calorimeter (where the mass of contained water is the same as the mass of the specimen).

Step 4: After the nails have given their thermal energy to the water, record the final temperature of the water in each calorimeter—which is the same as the final temperature of each specimen. Enter your findings in Table 1.

Table 1

Specimen	Mass (g)	$T_{initial}$ (°C)	T_{final} (°C)	ΔT (°C)	c (cal/g·°C)

Summing Up

Compare your values for the specific heats of your specimens to the table below. How do your values compare?

_____

Going Further

Try an unknown specimen and see how closely it matches the value of one in the table of specific heat capacities (Table 2).

Table 2

Specific Heat Capacities	
Substance	c (cal/g·°C)
Aluminum	0.215
Copper	0.0923
Gold	0.0301
Lead	0.0305
Silver	0.0558
Tungsten	0.0321
Zinc	0.0925
Mercury	0.033
Water	1.00

Name _____ Date _____

| CONCEPTUAL INTEGRATED SCIENCE | Experiment |

Heat: Heat Transfer—Radiation

Canned Heat I

Purpose
In this experiment, you will compare the ability of different surfaces to absorb thermal radiation.

Required Equipment and Supplies
heat lamp and base
radiation cans: silver, black, and white
thermometer
access to cold tap water
paper towel
graph paper

Discussion
Does the color of a surface make a difference in how well it absorbs thermal radiation? If so, how? The answers to these questions could help you decide what to wear on a hot, sunny day and what color to paint your house if you live in a hot, sunny climate. In this experiment, we will compare the thermal absorption ability of three surfaces: silver, black, and white. We'll do this by filling cans with these surfaces with water, then exposing the cans to heat lamps. We'll measure the temperature of the water in each of the cans while they're being exposed to the heat and see if there's a difference in the rate at which the temperatures increase.

Procedure
1. In which of the three cans do you think the water will heat up at the fastest rate?

_____

2. In which of the three cans do you think the water will heat up at the slowest rate?

_____

Step 1: Arrange the apparatus so that the heat lamp will shine equally on all three cans. The cans should be about 1 foot in front of the lamp. Do not turn the light on yet.

Step 2: Fill the cans with cold water and wipe up any spills. Quickly measure the initial temperature of the water in each of the cans and record it in Table 1.

Step 3: Turn on the lamp and start timing. Place the thermometer in the silver can.

Step 4: At the 1-minute mark (1:00), read the temperature of the water in the silver can and record it in Table 1. Quickly move the thermometer to the black can. Gently swirl the water in the can with the thermometer.

Step 5: At the 2-minute mark (2:00), read the temperature of the water in the black can and record it in Table 1. Quickly move the thermometer to the white can. Gently swirl the water in the can with the thermometer.

Step 6: At the 3-minute mark (3:00), read the temperature of the water in the white can and record it in Table 1. Quickly move the thermometer to the silver can. Gently swirl the water in the can with the thermometer.

Step 7: Repeat Steps 4–6 until the 21-minute mark temperature reading is made.

Table 1

Silver Can Temperatures (°C)	Black Can Temperatures (°C)	White Can Temperatures (°C)
Initial	Initial	Initial
T at 1:00	T at 2:00	T at 3:00
T at 4:00	T at 5:00	T at 6:00
T at 7:00	T at 8:00	T at 9:00
T at 10:00	T at 11:00	T at 12:00
T at 13:00	T at 14:00	T at 15:00
T at 16:00	T at 17:00	T at 18:00
T at 19:00	T at 20:00	T at 21:00

Step 8: Turn off the heat lamp.

3. Plot your data for all three cans on a single temperature vs. time graph.

Step 9: Determine the change in temperature for the water in each can while the heat lamp was on.

4. Determine the temperature change in the silver can while the heat lamp was on. Subtract the 1-minute mark temperature reading from the 19-minute mark temperature reading.

   Temperature change in silver can: _____ °C

5. Determine the temperature change in the black can while the heat lamp was on. Subtract the 2-minute mark temperature reading from the 20-minute mark temperature reading.

   Temperature change in black can: _____ °C

6. Determine the temperature change in the white can while the heat lamp was on. Subtract the 3-minute mark temperature reading from the 21-minute mark temperature reading.

   Temperature change in white can: _____ °C

Summing Up

1. In which can did the water heat up at the fastest rate? In which can did the water heat up at the slowest rate? Did your observations match your predictions?

   _____

2. Which would be a better choice if you were going to spend a long time outdoors on a hot, sunny day: a black T-shirt or a white T-shirt?

   _____

3. What happens to the thermal radiation that falls on each of the cans: is it absorbed or reflected?

   Silver: _____   Black: _____   White: _____

| **CONCEPTUAL INTEGRATED SCIENCE** | **Experiment** |

Heat: Heat Transfer—Radiation

Canned Heat II

Purpose
In this experiment, you will compare the ability of different surfaces to radiate thermal energy.

Required Equipment and Supplies
radiation cans: silver and black
thermometer
access to hot water
paper towel
graph paper

Discussion
Does the color of a surface make a difference in how well it radiates thermal energy? If so, how? The answer to these questions could help you decide what color coffeepot will best keep its heat and what color might be used to radiate heat away from a computer chip. In this experiment, we will compare the thermal radiation ability of two surfaces: silver and black. We'll do this by filling cans with these surfaces with hot water, then letting the water cool down. We'll measure the temperature of the water in both cans while they're cooling down and see if there's a difference in the rate at which the temperatures decrease.

Procedure
1. In which of the two cans do you think the water will cool down at the faster rate?

_____

2. In which of the two cans do you think the water will cool down at the slower rate?

_____

Step 1: Carefully fill the cans with hot water and wipe up any spills. Quickly measure the initial temperature of the water in each of the cans and record it in Table 1. Place the thermometer in the silver can.

Step 2: At the 1-minute mark, read the temperature of the water in the silver can and record it in Table 1. Quickly move the thermometer to the black can. Gently swirl the water in the can with the thermometer.

Step 3: At the 2-minute mark, read the temperature of the water in the black can and record it in Table 1. Quickly move the thermometer to the white can. Gently swirl the water in the can with the thermometer.

Step 4: Repeat Steps 2–3 until the 20-minute mark temperature reading is made.
3. Plot your data for both cans on a single temperature vs. time graph.

Table 1

Silver Can Temperatures (°C)	Black Can Temperatures (°C)
Initial	Initial
T at 1:00	T at 2:00
T at 3:00	T at 4:00
T at 5:00	T at 6:00
T at 7:00	T at 8:00
T at 9:00	T at 10:00
T at 11:00	T at 12:00
T at 13:00	T at 14:00
T at 15:00	T at 16:00
T at 17:00	T at 18:00
T at 19:00	T at 20:00

Step 5: Determine the change in temperature for the water in each can while it was allowed to cool.

4. Determine the temperature change in the silver can. Subtract the 1-minute mark temperature reading from the 19-minute mark temperature reading.

   Temperature change in silver can: _____ °C

5. Determine the temperature change in the black can. Subtract the 2-minute mark temperature reading from the 20-minute mark temperature reading.

   Temperature change in black can: _____ °C

Summing Up

1. In which can did the water cool down faster? In which can did the water cool down more slowly? Did your observations match your predictions?

   _____

2. What color should the surface of a coffeepot be in order to keep the hot water inside it hot for the longest time?

   _____

3. The central processing units (CPUs) in personal computers sometimes get very hot. To prevent damage and improve processing speed, they need to be kept cool. Often, a piece of metal is placed in contact with the CPU to draw some heat away. The metal then radiates its heat to the surroundings. To best remove heat from the CPU, should the metal's surface be colored black or left silvery?

   _____

Laboratory Manual for *Conceptual Integrated Science,* © 2007 Addison Wesley

Name _____ Date _____

Heat: Heat Transfer—Conduction and Radiation
I'm Melting! I'm Melting!

Purpose
In this activity, you will investigate the curious heat transfer ability of different surfaces.

Required Equipment and Supplies
defrosting tray (Miracle Thaw® or equivalent)
a second defrosting tray wrapped tightly in aluminum foil
white Styrofoam plate
black Styrofoam plate (white plate colored completely using a felt-tip pen)
four ice cubes of similar size
paper towel

Discussion
If you walk around the house with bare feet, you probably notice that a tile floor feels much colder than a carpeted floor or rug. It's hard to believe that they might actually have the same temperature. The tile feels colder because it is a better conductor than carpet. Heat is conducted from your warmer feet to the cooler floor faster when the floor is tile than when the floor is carpet. So your feet are cooled faster by tile than they are by carpet at the same cool temperature. In this activity, you will see which kinds of surfaces transfer heat most rapidly.

Procedure
Step 1: Set the defrosting tray, defrosting tray covered in aluminum foil, white Styrofoam plate, and blackened Styrofoam plate out on your table.

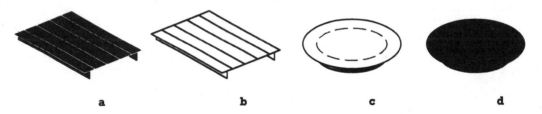

a   b   c   d

Figure 1. a. defrosting tray. b. defrosting tray covered in aluminum foil. c. white Styrofoam plate. d. blackened Styrofoam plate.

1. Which surfaces feel colder and which ones feel warmer? (Just touch a corner; don't transfer too much of your own body heat to any of the objects.)

_____

Step 2: In a moment, you will set an ice cube on each of the surfaces. When you do, they will begin melting. Before you set the ice cubes out, make some predictions.

2. Which ice cube will melt most quickly?

_____

3. Which ice cube will melt most slowly?

_____

Step 3: Set the ice cubes out on their respective surfaces quickly. Observe the ice cubes for several minutes (preferably until the fastest-melting ice cube melts completely).

4. Which ice cube melted most quickly?

_____

5. Which ice cube is melting most slowly?

_____

Summing Up

1. How do your observations compare to your predictions?

_____

_____

2. Which way did the heat flow in this activity? (From what to what?)

_____

_____

3. Two of the surfaces were conductors and two of the surfaces were insulators. Which were which?

_____

_____

4. What advantage—in terms of heat transfer—did one defrosting tray have over the other?

_____

_____

5. What advantage—in terms of heat transfer—did one Styrofoam plate have over the other?

_____

_____

6. Which of the following conclusions are supported by your observations and which are not? Give evidence from this activity to justify your answer.

   a. "Metals transfer heat faster than Styrofoam."

   _____

   _____

   b. "Black surfaces transfer heat faster than non-black surfaces."

   _____

   _____

Name _____ Date _____

Electricity and Magnetism: Electric Force and Charge
A Force to Be Reckoned

Purpose
In this investigation, you will explore the nature of a force. You will determine whether or not the force is distinct from other known forces.

Required Equipment and Supplies
a pith ball suspended from a support rod
vinyl and acetate strips or tubes
wool and silk cloth swatches (2″ × 2″)
brick (or a heavy book)
bar magnet

Procedure
Step 1: Generate an attractive force. Rub the vinyl with the wool. Hold the vinyl near the pith ball and see if the strip will attract the pith ball to it as shown. If it won't, try rubbing the vinyl more vigorously, or try using the acetate rubbed with silk. If you continue to have difficulty, ask your teacher for assistance.

Once you have seen the plastic strip (vinyl or acetate) attract the pith ball, set the plastic aside and discharge the pith ball by gently touching it.

Consider the suggestion that the attraction you observed is simply the result of gravitational force. After all, *gravitational force* causes any two things with mass to be attracted to each other. Recall that the force of attraction is proportional to the amount of mass involved.

Step 2: Hold the brick close to the pith ball. The mass of the brick is much greater than the mass of the plastic strip. What—if anything—does the brick do to the pith ball? What does this tell you about the suggestion that the attraction between the plastic strip and the pith ball is *gravitational?*

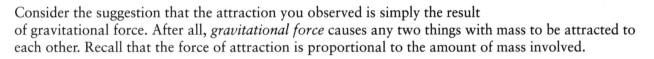

Step 3: Generate a repulsive force. Using the plastic strip (rubbed with the cloth as was done in Step 1), try to repel the pith ball as shown. Does the observation of repulsion support or contradict the suggestion that the attraction between the plastic strip and the pith ball is gravitational? Why?

_____

_____

Once you have seen the plastic strip (vinyl or acetate) attract the pith ball, set the plastic aside and discharge the pith ball by touching it.

Consider the suggestion that the attraction and repulsion you observed are simply the result of *magnetic force*. After all, magnets can both attract *and* repel.

Step 4: Hold the bar magnet close to the pith ball. What does the magnet do to the pith ball, and what does this tell you about the suggestion that the attraction and repulsion between the plastic strip and the pith ball are magnetic?

_____

_____

Summing Up
The force between the plastic strip and the pith ball is different from gravitational force and different from magnetic force. It is called *electrostatic force*, and it is a force between any two objects that are electrically charged.

1. When the vinyl is rubbed with wool, the vinyl gets a *negative* charge. What kind of charge is on the pith ball when the charged vinyl repels it?

_____

2. Does electrostatic force get stronger or weaker with distance? Does the interaction become stronger when charged objects get closer together or when they get farther apart?

_____

_____

Laboratory Manual for *Conceptual Integrated Science,* © 2007 Addison Wesley

CONCEPTUAL INTEGRATED SCIENCE	Activity

Electricity and Magnetism: Conductors and Insulators
Charging Ahead

Purpose
In this activity, you will observe the effects and behavior of static electricity.

Required Equipment and Supplies
two new balloons
Van de Graaff generator
several pie tins (small)
several Styrofoam bowls
bubble-making materials (solution and wand)
matches

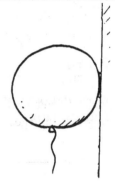

Discussion
Scuff your feet across a rug and reach for a doorknob and zap—electric shock! The electrical charge that makes up the spark can be several thousand volts, which is why technicians have to be so careful when working with tiny circuits such as those in computer chips!

Procedure
Step 1: Blow up a balloon. After stroking it against your hair, place it near some small pieces of Styrofoam or puffed rice. Then place the balloon against the wall where it will "stick," as shown to the right. On the drawing, sketch the arrangement of some sample charges on the balloon and on the wall.

Step 2: Blow up a second balloon. Rub both balloons against your hair. Do they attract or repel each other?

_____

Step 3: Stack several pie tins on the dome of the Van de Graaff generator. Turn the generator on. What happens and why?

_____

_____

_____

Step 4: Turn the generator off and discharge it with the discharge ball or by touching it with your knuckle. Stack several Styrofoam bowls in the generator and turn the generator on. What happens and why?

_____

_____

_____

Step 5: With the Van de Graaff generator off and discharged, blow some bubbles toward it. Observe the behavior of the bubbles. Then turn the generator on and blow bubbles toward it again. Watch carefully. What happens?

_____

_____

_____

Step 6: Stand on an isolation stand (or rubber mat) next to a discharged Van de Graaff generator. Place one hand on the conducting sphere on top of the generator and have your partner switch on the generator motor. Shake your head as the generator charges up. What do you experience?

_____

_____

_____

Summing Up

Which of the demonstrations in this activity are better explained by the principle that like charges repel and opposites attract? Which are better explained in terms of the differences between conductors and insulators?

_____

_____

Going Further

Light a wooden match and move it near a charged sphere on top of the generator. What happens and why?

_____

_____

Laboratory Manual for _Conceptual Integrated Science,_ © 2007 Addison Wesley

Name _____ Date _____

Electricity and Magnetism: Ohm's Law

Ohm, Ohm on the Range

Purpose

In this experiment, you will arrange a simple circuit involving a power source and a resistor. You will attach an ammeter and a voltmeter to the circuit. You will measure corresponding values of current and voltage in the circuit. You will then interpret your observations to find the relationship between current, voltage, and resistance.

Required Equipment and Supplies

variable DC power supply (0–6 V)
2 power resistors with different resistances (values between 3 Ω and 10 Ω recommended)
power resistor with unknown resistance (for the Going Further section of experiment)
miniature lightbulb in socket (14.4-V flashlight bulb recommended)
DC ammeter (0–1 A analog recommended)
DC voltmeter (0–10 V analog recommended)
5 connecting wires
graph paper

Discussion

The current, voltage, and resistance in an electric circuit are related to one another in a very specific way. Designers of electric circuits must take this relationship into account or their circuits will fail. This relationship is as important and fundamental in electricity as Newton's second law of motion is in mechanics. In this experiment, you will determine this relationship.

Procedure

Part A: Connecting the Meters

Step 1: With the power supply turned down to zero, arrange a simple circuit using the power supply, the miniature bulb, and two connecting wires as shown in Figure 1. If your power supply has "AC" terminals, do not use them; connect only to the "DC" terminals.

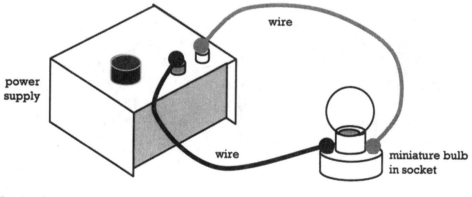

Figure 1

Step 2: Verify that the circuit is working properly by slowly turning the knob on the power supply to increase the power to the circuit. The bulb should begin to glow and increase in brightness as power is increased. If the circuit doesn't work, make adjustments so that it does. Ask your instructor for assistance if necessary.

Step 3: Turn the power down to zero.

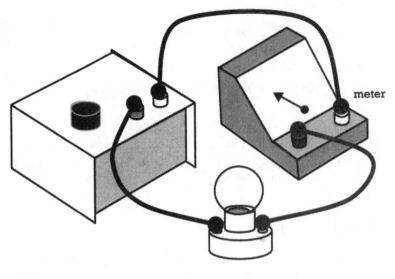

Figure 2

Step 4: Connect the ammeter in series as shown in Figure 2. Slowly increase the power to the circuit and observe the ammeter and the bulb. If the needle on your ammeter ever moves in the wrong direction (that is, tries to "go negative"), reverse the connections and try again.

1. When the connections are correct, what do you observe?

_____

_____

_____

Step 5: Turn the power back down to zero. Disconnect the ammeter and restore the circuit to its original configuration (shown in Figure 1).

Step 6: Connect the ammeter in parallel as shown in Figure 3. Slowly increase the power to the circuit and observe the ammeter and the bulb.

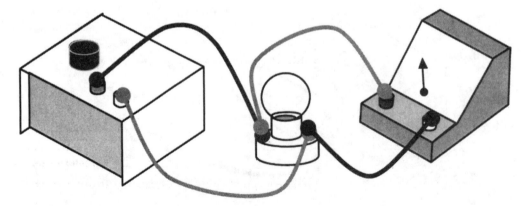

Figure 3

Laboratory Manual for *Conceptual Integrated Science,* © 2007 Addison Wesley

2. What do you observe?

_____

_____

_____

Step 7: Turn the power back down to zero. Disconnect the ammeter, and restore the circuit to its original configuration (shown in Figure 1).

Step 8: Connect the voltmeter in series as shown in Figure 2. Slowly increase the power to the circuit and observe the voltmeter and the bulb. If the needle on your voltmeter ever moves in the wrong direction (tries to "go negative"), reverse the connections and try again.

3. When the connections are correct, what do you observe?

_____

_____

_____

Step 9: Turn the power back down to zero. Disconnect the voltmeter and restore the circuit to its original configuration (shown in Figure 1).

Step 10: Connect the voltmeter in parallel as shown in Figure 2. Slowly increase the power to the circuit and observe the voltmeter and the bulb.

4. What do you observe?

_____

_____

_____

When the ammeter is connected correctly, the circuit behaves as it did when no meters were connected (increasing the power increases the brightness of the bulb). When the power is increased, the ammeter shows increased current in the circuit. When the ammeter is connected incorrectly, the bulb remains dim or does not light at all, although the ammeter shows significant current.

5. The correct method is to connect the ammeter to the circuit in ___series or ___parallel (*select one*).

When the voltmeter is connected correctly, the circuit behaves as it did when no meters were connected (increasing the power increases the brightness of the bulb). When the power is increased, the voltmeter shows increased current in the circuit. When the voltmeter is connected incorrectly, the bulb remains dim or does not light at all, although the voltmeter shows significant voltage.

6. The correct method is to connect the voltmeter to the circuit in ___series or ___parallel (*select one*).

Step 11: Connect a circuit that includes the bulb, the ammeter, and the voltmeter. Connect the ammeter and voltmeter correctly, based on your findings. Sketch a diagram in the space below to show how the wires connect the various circuit elements.

Step 12: Verify that the circuit is working correctly. When the power is increased, the bulb should get brighter, the ammeter reading should increase, and the voltmeter reading should increase. Show your instructor your working circuit.

Part B: Collecting Data

Step 1: Turn the power down to zero. Replace the bulb with one of the known resistors. (Remove the bulb and socket from the circuit; connect the resistor in its place.)

Step 2: Increase the power until the current indicated on the ammeter is 0.10 A.

Step 3: Record the corresponding voltmeter reading in the appropriate space in the data table.

Step 4: Repeat Steps 2 and 3 for current values of 0.20 A through 0.60 A.

Step 5: Turn the power down to zero. Replace the first resistor with the second resistor.

Step 6: Repeat Steps 2 through 4 to complete the data table for the second resistor.

Data

Current I (amperes)	First Known Resistor Voltage V (volts)	Second Known Resistor Voltage V (volts)
0	0	0
0.10		
0.20		
0.30		
0.40		
0.50		
0.60		

Step 7: On your graph paper, plot graphs of voltage versus current for all three resistors. Plot all three data sets on one set of axes. Voltage will be the vertical axis; current will be the horizontal axis. Scale the graph to accommodate all your data points. Label each axis as to its quantity and units of measurement.

Step 8: Make a line of best fit for each data set plotted on the graph.

Step 9: Determine the slope of each best-fit line.

1. What is the slope of each best-fit line? Identify each slope by its corresponding resistor value. Don't forget to include the correct units for each slope value.

   _____-Ω resistor best-fit line slope = _____

   _____-Ω resistor best-fit line slope = _____

   _____-Ω resistor best-fit line slope = _____

Observe the similarity between the values of the slope and the values of the corresponding resistance.

Going Further

Obtain a power resistor with an unknown resistance from your instructor. Using the techniques of this experiment, determine the resistance of the resistor. Record your data and calculations in the space below.

Once you determine the resistance of the unknown resistor, ask your instructor for the accepted value. Record that value and calculate the percent error in your value.

Summing Up

1. Which mathematical expression shows the correct relationship between current, voltage, and resistance?

    ___R = IV   ___R = I/V   ___R = V/I

The correct answer is one form of the equation known as Ohm's Law.

2. Suppose voltage versus current data were taken for two devices, A and B, and the results were plotted to form the best-fit lines shown in Figure 4. Which device has the greater resistance? How do you know?

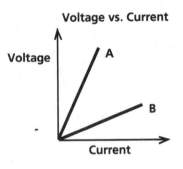

Voltage vs. Current

_____

_____ **Figure 4**

_____

_____

3. Suppose the voltage and current data for electrical device C produced graph C shown in Figure 5. How does the resistance of device C change as the current increases? Explain.

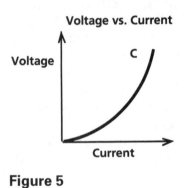

Voltage vs. Current

_____

_____ **Figure 5**

_____

_____

4. Suppose the voltage and current data for electrical device D produced graph D shown in Figure 6. How does the resistance of device D change as the current increases? Explain.

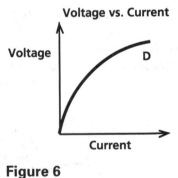
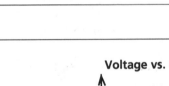

Voltage vs. Current

_____

_____ **Figure 6**

_____

_____

        Laboratory Manual for *Conceptual Integrated Science,* © 2007 Addison Wesley

Name _____ Date _____

Electricity and Magnetism: Electric Circuits

Batteries and Bulbs

Purpose
In this activity, you will explore various arrangements of batteries and bulbs and the effects of those arrangements on bulb brightness.

Required Equipment and Supplies
2 D-cell batteries
4 connecting wires
2 miniature bulbs (1.5-volt or 2.5-volt flashlight bulbs)
2 miniature bulb sockets

Discussion
Many devices include electronic circuitry, most of which are quite complicated. Complex circuits are made, however, from simple circuits. In this activity you will build one of the simplest yet most useful circuits ever invented—that for lighting a lightbulb!

Procedure
Step 1: Remove the bulb from the mini socket.
1. On a separate sheet of paper, draw a detailed diagram of the bulb, showing the following parts of the bulb's "anatomy."
   - glass bulb
   - filament leads (tiny wires that lead to the filament)
   - screw base
   - base contact (made of lead)
   - lead separator (glass bead)

2. There are four parts of the bulb's anatomy that you can touch (without having to break the bulb). Two of them are made of *conducting* material (metal), and two are made of *insulating* material. List them.

   Conducting parts on the outside of the bulb:_____

   Insulating parts on the outside of the bulb:_____

Step 2: Examine the two diagrams of a working electric circuit shown below. The diagram on the left shows pictorial representations of circuit elements. The diagram on the right shows symbolic representations. Use the symbolic representations in the steps that follow.

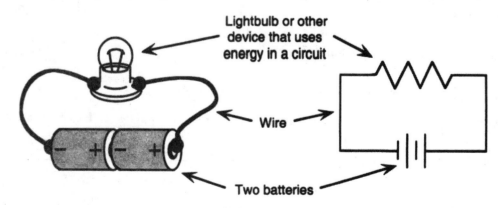

Using a bare bulb (out of its socket), one battery, and **two** wires, try lighting the bulb in as many ways as you can. On a separate sheet of paper, sketch at least two *different* arrangements that work. Also sketch at least two arrangements that don't work. Be sure to label them as "works" or "doesn't work."

Step 3: Using a bare bulb (out of its socket), one battery, and **one** wire, try lighting the bulb as many ways as you can. Sketch your arrangements, and note the ones that work.

3. Is it possible to light the bulb using the battery and **no** wires? Explain.

_____

Step 4: Connect one bulb (in its socket) to two batteries as shown in Figure 1. This arrangement is often referred to as a *simple* circuit.

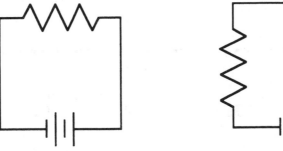

Figure 1. Simple circuit.   **Figure 2.** Series circuit.

Step 5: Connect the bulbs, batteries, and wire as shown in Figure 2. When the bulbs are connected one after the other like this, the result is called a *series* circuit.

4. How does the brightness of each bulb in the series circuit compare to the brightness of the bulb in a simple circuit?

_____

5. What happens if one of the bulbs in a series circuit is removed? (Do this by unscrewing a bulb from its socket.)

_____

Step 6: Connect the bulbs, batteries, and wire as shown in Figure 3. When the bulbs are connected along separate paths like this, the result is called a *parallel* circuit.

6. How does the brightness of each bulb in the parallel circuit compare to the brightness of the bulb in a simple circuit?

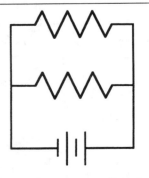

_____

7. What happens if one of the bulbs in a parallel circuit is removed?

Figure 3. Parallel circuit.

_____

Summing Up

1. With what two parts of the bulb does the bulb socket make contact?

_____

_____

2. What do successful arrangements of batteries and bulbs have in common?

_____

_____

3. How do you suppose most of the circuits in your home are wired—in series or in parallel? What is your evidence?

_____

_____

4. How do you suppose automobile headlights are wired—in series or in parallel? What is your evidence?

_____

_____

Name _____ Date _____

Electricity and Magnetism: Electric Circuits

An Open and Short Case

Purpose
In this activity, you will explore open circuits and short circuits.

Required Equipment and Supplies
2 D-cell batteries
5 connecting wires
2 miniature bulbs (1.5-volt or 2.5-volt flashlight bulbs)
2 miniature bulb sockets
DC ammeter (0–5A)

Discussion
Electrical circuits are all around us. (We often appreciate them most when the power goes out.) Most circuits work perfectly well (when power is available). But electrical circuits *can* fail. Two common modes of circuit failure are *open circuits* and *short circuits*. In this activity, we will learn how these circuit failures are similar and how they are different.

We will be using an *ammeter* to help us with this investigation. An ammeter is a simple device used to measure electric current—the rate at which charge flows through a circuit. Current is measured in amperes ("amps").

Procedure
Part A: Open and Short Circuits
Step 1: Arrange a simple circuit using two batteries, a bulb, an ammeter, and three connecting wires as shown in Figure 1. If the circuit is working, the bulb will light, and some amount of current will register on the ammeter. It should be less than 1 amp.

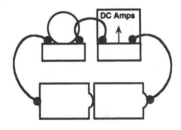

Figure 1. Simple circuit.

Step 2: Predict what would happen to the simple circuit if one of the wires were disconnected at one point in the circuit. (Don't touch the circuit yet—predict first!)

1. What will happen to the bulb and what will happen to the reading on the ammeter (compared to what happened in Step 1)?

   _____

   _____

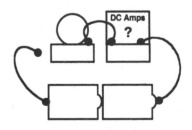

Figure 2. Open circuit.

Once you've made your prediction and discussed it with your partner(s), disconnect a wire as shown in Figure 2. This is an *open circuit*.

2. Record your observations.

   _____

   _____

Step 3: Predict what would happen to the simple circuit if an additional wire were added to the circuit so as to connect the terminals of the bulb to each other as shown in Figure 3. (Don't touch the circuit yet—predict first!)

3. What will happen to the bulb and what will happen to the reading on the ammeter (compared to what happened in Step 1)?

_____

_____

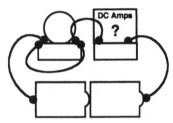

Once you've made your prediction and discussed it with your partner(s), add a wire as shown in Figure 3. This is a *short circuit*.

Figure 3. Short circuit.

4. Record your observations.

_____

_____

Step 4: One of these circuit failures is said to have almost *no electrical resistance* and one is said to have *infinite electrical resistance*. Electrical resistance is inversely proportional to electrical current in a simple circuit.

5. Which circuit failure has no current and therefore infinite electrical resistance?

_____

6. Which circuit failure has a large amount of current and therefore almost no electrical resistance?

_____

Part B: Short-Circuited Series Circuit

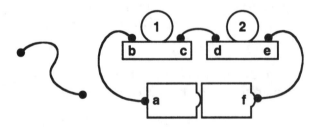

Figure 4. Series circuit and additional wire.

Step 5: Arrange the series circuit shown in Figure 4. Notice that bulbs 1 and 2 light up. Notice there is an additional wire not yet in the circuit.

Step 6: Add the additional wire to the circuit, connecting point a to point b. Notice that both bulb 1 and bulb 2 remain fully lit.

Step 7: Now use the additional wire to connect point b to point c. Notice that bulb 1 goes out (or becomes much dimmer), while bulb 2 remains fully lit (or becomes brighter).

Laboratory Manual for *Conceptual Integrated Science,* © 2007 Addison Wesley

7. Predict what will happen if the additional wire is used to connect other points on the circuit. Make your prediction in terms of what will happen to each of the bulbs. Will bulb 1 remain lit or go out? Will bulb 2 remain lit or go out?

Important Note: For purposes of this activity, "remaining lit" includes increased brightness, and "going out" includes significant dimming.

Remember to make predictions before making observations!

a. If point c is connected to point d, bulb 1 will __remain lit __go out, and bulb 2 will __remain lit __go out.

b. If point d is connected to point e, _____.

c. If point e is connected to point f, _____.

d. If point f is connected to point a, _____.

e. If point a is connected to point c, _____.

f. If point a is connected to point d, _____.

g. If point a is connected to point e, _____.

h. If point b is connected to point d, _____.

i. If point b is connected to point e, _____.

Step 8: Observe what happens if the additional wire is used to make each of the connections.

a. When point c is connected to point d, bulb 1 __remains lit __goes out, and bulb 2 __remains lit __goes out. (*Check the correct answers*)

b. When point d is connected to point e, _____.

c. When point e is connected to point f, _____.

d. When point f is connected to point a, _____.

e. When point a is connected to point c, _____.

f. When point a is connected to point d, _____.

g. When point a is connected to point e, _____.

h. When point b is connected to point d, _____.

i. When point b is connected to point e, _____.

Part C: Short-Circuited Parallel Circuit

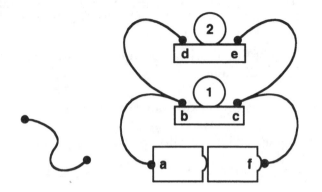

Figure 5. Parallel circuit and additional wire.

Step 9: Arrange the parallel circuit shown in Figure 5. Notice that bulbs 1 and 2 light up. Notice there is an additional wire not yet in the circuit.

Step 10: Add the additional wire to the circuit, connecting point a to point b. Notice that both bulb 1 and bulb 2 remain fully lit.

Step 11: Now use the additional wire to connect point b to point c. Notice that bulb 1 goes out (or becomes much dimmer), while bulb 2 remains fully lit (or becomes brighter).

8. Predict what will happen if the additional wire is used to connect other points on the circuit. Make your prediction in terms of what will happen to each of the bulbs. Will bulb 1 remain lit or go out? Will bulb 2 remain lit or go out? **For purposes of this activity, "remaining lit" includes increased brightness, and "going out" includes significant dimming.** Remember to make predictions before making observations!

   a. If point c is connected to point d, bulb 1 will __remain lit __go out, and bulb 2 will __remain lit __go out.

   b. If point d is connected to point e, _____

   c. If point e is connected to point f, _____

   d. If point f is connected to point a, _____

   e. If point a is connected to point c, _____

   f. If point a is connected to point d, _____

   g. If point a is connected to point e, _____

   h. If point b is connected to point d, _____

   i. If point b is connected to point e, _____

    Laboratory Manual for *Conceptual Integrated Science,* © 2007 Addison Wesley

Step 12: Observe what happens if the additional wire is used to make each of the connections.

a. When point c is connected to point d, bulb 1 __remains lit __goes out, and bulb 2 __remains lit __goes out.

b. When point d is connected to point e, _____

c. When point e is connected to point f, _____

d. When point f is connected to point a, _____

e. When point a is connected to point c, _____

f. When point a is connected to point d, _____

g. When point a is connected to point e, _____

h. When point b is connected to point d, _____

i. When point b is connected to point e, _____

Summing Up

1. What do open circuits and short circuits have in common?

_____

_____

2. How are open circuits and short circuits different?

_____

_____

3. Examine the cases in Parts B and C when both bulbs went out. Is there anything that *all* those cases have in common?

_____

_____

CONCEPTUAL INTEGRATED SCIENCE	Activity

Electricity and Magnetism: Electric Circuits

Be the Battery

Purpose
In this activity, you will provide energy to an electric circuit using your own muscle power.

Required Equipment and Supplies
Genecon® handheld generator (or equivalent)
3 miniature bulbs (6-volt flashlight bulbs)
3 miniature bulb sockets
6 connecting wires

Discussion
Batteries last longer in some circuits than they do in others. They last longer when they don't have to "work" so hard. In this activity, you will do the work of the battery. That is, *you* will power a circuit using the handheld generator. You will learn which circuits are easier to power and which circuits are harder to power. And you'll gain a better appreciation for what batteries and the local power utility do for you all the time!

Procedure
Step 1: Arrange a simple circuit using the generator and a bulb in its socket as shown in Figure 1. Gently crank the handle to make the bulb light up. Take care not to crank the generator too quickly, and don't give it any sudden jerks or bursts of motion.

Figure 1. The hand powered circuit.

1. When the bulb is lit, how can you make it brighter? Does this require more effort on your part?

_____

Step 2: While cranking the generator and lighting the bulb, have a partner unscrew the bulb from the socket as shown in Figure 2.

2. What happens to the cranking effort when the bulb is unscrewed from its socket?

_____

3. When the bulb is removed from the socket, is the resulting circuit an open circuit or a short circuit?

_____

Figure 2

4. Is the electrical resistance in this kind of circuit very high or very low?

_____

Step 3: Remove the generator leads from the bulb terminals and connect them to each other as shown in Figure 3.

5. What happens to the cranking effort when the generator leads are connected to each other?

_____

6. When the generator leads are connected directly to each other, is the resulting circuit an open circuit or a short circuit?

_____

7. Is the electrical resistance in this kind of circuit very high or very low?

_____

Figure 3

Step 4: Connect three bulbs in series as shown in Figure 4. Gently crank the handle to make the bulb light up. Get a sense of how much effort is needed to power the circuit.

Figure 4. A hand-powered series circuit.

Step 5: Connect three bulbs in parallel as shown in Figure 5. Gently crank the handle to make the bulb light up. Get a sense of how much effort is needed to power the circuit.

Figure 5. A hand-powered parallel circuit.

8. Which circuit is harder to power, the series circuit or the parallel circuit?

_____

Laboratory Manual for *Conceptual Integrated Science,* © 2007 Addison Wesley

Summing Up

1. Which types of circuits are harder to power, those having low electrical resistance or those having high electrical resistance?

_____

2. Which arrangement of three bulbs has more electrical resistance, series or parallel? Justify your answer using observations from the activity.

_____

_____

3. Under which conditions will a battery run down faster, when connected to a high-resistance circuit or when connected to a low-resistance circuit?

_____

4. Which battery would last longer, one connected to a three-bulb series circuit or one connected to a three-bulb parallel circuit? (Assume the batteries are identical and the bulbs are identical.)

_____

CONCEPTUAL INTEGRATED SCIENCE	Activity

Electricity and Magnetism: Magnetic Fields

Magnetic Personality

Purpose
In this activity, you will explore the patterns of magnetic fields around bar magnets in various configurations.

Required Equipment and Supplies
3 bar magnets
iron filings and paper or
a magnetic field projectual (iron filings suspended in oil encased in an acrylic envelope)

Discussion
A magnetic field is a kind of aura that surrounds magnets. Although it can't be seen directly, the overall shape of the field can be seen by the effect it has on iron filings.

Procedure
Step 1: Place a bar magnet on a horizontal surface such as your tabletop. Use the iron filings to see the pattern of the magnetic field.

Iron Filings and Paper Method
Cover the magnet or magnets with a sheet of paper. Then sprinkle iron filings on top of the paper. Jiggle the paper a little bit to help the iron filings find their way into the magnetic field pattern.

Projectual Method
Mix the iron filings by rotating the projectual. Use the glass rod inside the projectual to help stir the iron filings into a fairly even distribution. Hold the projectual upside down for several seconds before placing it on the magnet or magnets. Take care not to scratch the surface of the projectual by moving it across the magnets once it is in place.

Sketch the field for a single bar magnet in Figure 1.

Step 2: Arrange two bar magnets in a line with opposite poles facing each other. Leave about an inch between the poles. Use the iron filings to see the pattern of the magnetic field. Sketch the field for opposite poles in Figure 2.

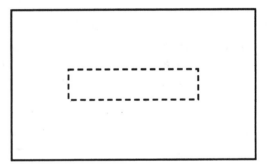

Figure 1

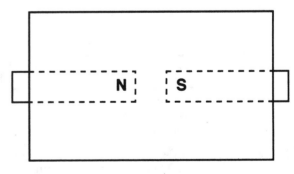

Figure 2

Step 3: Arrange two bar magnets in a line with north poles facing each other. Leave about an inch between the poles. Use the iron filings to reveal the magnetic field. Sketch the field for north poles in Figure 3.

Step 4: Predict the pattern for the magnetic field of two south poles facing each other as shown in Figure 4. Make a predictive sketch in Figure 4.a. After completing your prediction, arrange two bar magnets in a line with south poles facing each other. Leave about an inch between the poles. Use the iron filings to reveal the magnetic field. Sketch the field for south poles in Figure 4.b.

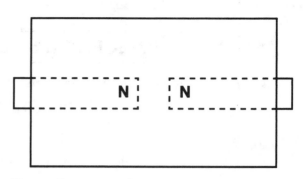

Figure 3

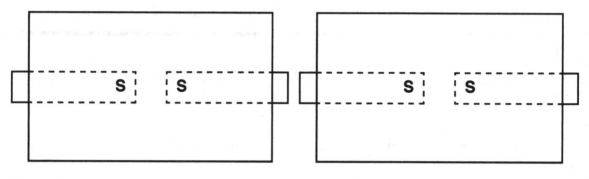

Figure 4.a   **Figure 4.b**

Step 5: Predict the pattern for the magnetic field of two bar magnets parallel to each other as shown in Figure 5. Make a sketch in Figure 5.a. After completing your prediction, arrange two bar magnets parallel to each other. Use the iron filings to reveal the magnetic field. Sketch the field for two magnets parallel to each other in Figure 5.b.

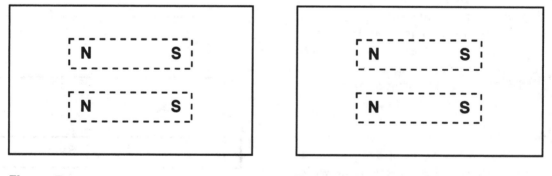

Figure 5.a   **Figure 5.b**

Step 6: Predict the pattern for the magnetic field of two bar magnets antiparallel to each other as shown in Figure 6. Make a sketch in Figure 6.a. After completing your prediction, arrange two bar magnets antiparallel to each other. Use the iron filings to reveal the magnetic field. Sketch the field for two magnets parallel to each other in Figure 6.b.

Laboratory Manual for *Conceptual Integrated Science,* © 2007 Addison Wesley

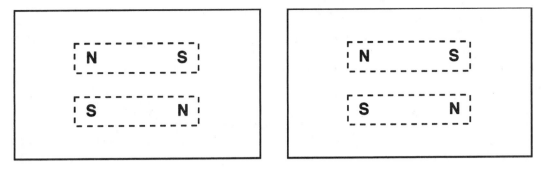

Figure 6.a   **Figure 6.b**

Step 7: Can you arrange three bar magnets to create the magnetic field shown in Figure 7? If so, how? Is there more than one way to do it?

_____

_____

_____

1. Does this pattern show attraction or repulsion?

_____

Step 8: Can you arrange three bar magnets to create the magnetic field shown in Figure 8? If so, how? Is there more than one way to do it?

_____

_____

_____

2. Does this pattern show attraction or repulsion?

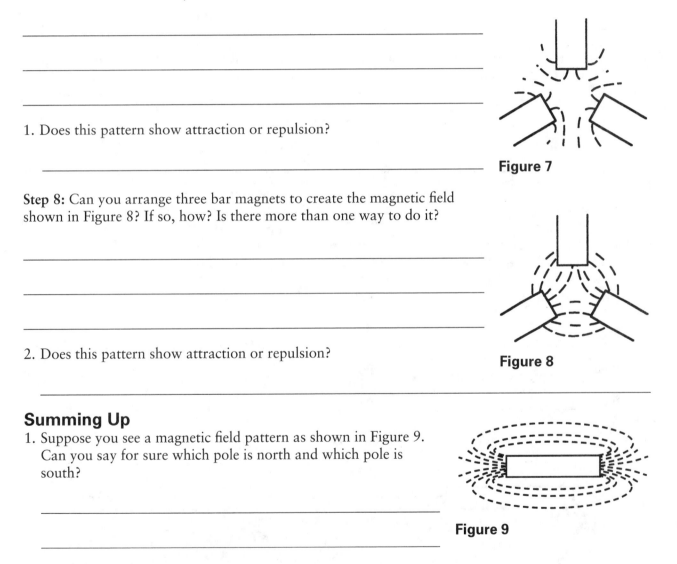

Figure 7

Figure 8

Summing Up

1. Suppose you see a magnetic field pattern as shown in Figure 9. Can you say for sure which pole is north and which pole is south?

_____

_____

2. Suppose you see a magnetic field pattern as shown in Figure 10. Can you say for sure which pole is north and which pole is south?

_____

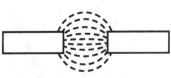

Figure 9

Figure 10

3. Suppose you see a magnetic field pattern as shown in Figure 11. If pole A is a north pole, what is pole B?

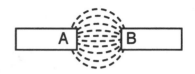

Figure 11

_____

_____

4. Suppose you see a magnetic field pattern as shown in Figure 12. If pole C is a north pole, what is pole D?

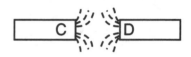

Figure 12

_____

_____

5. Suppose you see a magnetic field pattern as shown in Figure 13. If pole E is a north pole, what are poles F, G, H, I, J, K, and L?

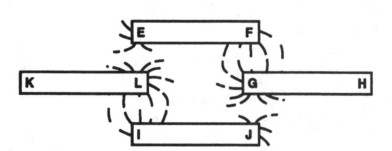

Figure 13

Pole E: North   Pole F:_____

Pole G:_____   Pole H:_____

Pole I:_____   Pole J:_____

Pole K:_____   Pole L:_____

6. Which of the patterns in Figure 14—if either—is/are possible using three bar magnets?

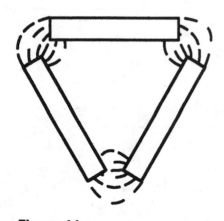   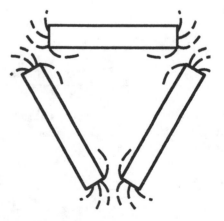

Figure 14.a   **Figure 14.b**

Name _____ Date _____

Electricity and Magnetism: Magnetic Fields

Electric Magnetism

Purpose
In this activity, you will investigate the electric origin of magnetic fields.

Required Equipment and Supplies
large battery (6 V lantern battery or 1.5 V ignitor battery)
4 small compasses
small platform (a discarded compact disc or equivalent)
support rod with base
ring clamp
connecting wires

Discussion
In Magnetic Personality, you investigated the magnetic field around various configurations of bar magnets. But where does the magnetic field come from? What's going on inside a bar magnet to make it magnetic? In this activity, you will discover the origin of all magnetic fields.

Procedure

Part A: Current Across a Compass
Step 1: Set a compass on your desktop and allow the needle to settle into its north–south alignment as shown in Figure 1.

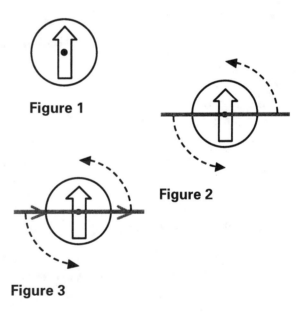

Figure 1

Figure 2

Step 2: Stretch a connecting wire across the top of the compass as shown in Figure 2. Rotate the wire clockwise and counterclockwise so that you can see that the wire itself has no effect on the compass needle.

Step 3: Connect the stretched wire to the battery to form a short circuit and again rotate the wire back and forth as shown in Figure 3. Keep the short circuit connected for no more than 10 seconds.

Figure 3

What effect does the current-carrying wire have on the compass?

_____

Step 4: Determine which is more effective in deflecting the compass needle: north–south current or east–west current. Keep in mind that short circuits must not be allowed to run more than 10 seconds, and compasses must be level to work properly.

Current has the greatest effect on the compass needle when it runs

___ north–south.   ___ east–west.   (*check one*)

Step 5: Try placing the wire **below** the compass and then running current through it.

Does the current affect the needle when the wire passes below the compass?

_____

Step 6: Try reversing the direction of the current by reversing the connections to the battery.

What difference does reversing the direction of current have on the deflection of the needle?

_____

Part B: Current Through a Platform of Compasses

Step 1: Arrange the apparatus as shown in Figure 4. Connecting wire passes through the center of the platform. The platform is supported by the ring clamp. The compasses are placed on the platform. Devise a method to have the connecting wire as vertical as possible as it passes through the compass platform.

Step 2: Before running any current through the wire, examine the compass needles by looking down from above the platform. Notice they all point north as indicated in Figure 5.

Step 3: Arrange to have current passing upward through the platform as shown in Figure 6. Connect the wire to the battery and tap the platform a few times. Record the new orientations of the compass needles in Figure 6.

Step 4: Reverse the direction of the current so the current passes downward through the platform. Connect the wire to the battery and tap the platform a few times. Record the new orientations of the compass needles in Figure 7.

The ability of an electric current to affect a compass needle was discovered by Hans Christian Ørsted, a Dutch high school teacher, in 1820. The observation established the connection between electricity and magnetism. We now know that **all** magnetic fields are the result of moving electric charge (even the magnetic fields of bar magnets). Ørsted's discovery stands as one of the most significant discoveries in the history of physics.

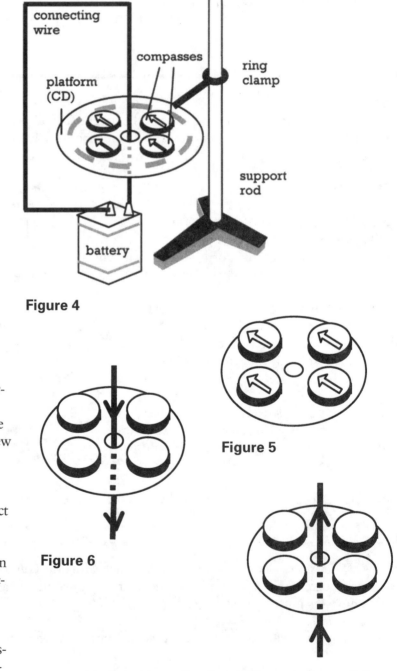

Figure 4

Figure 5

Figure 6

Figure 7

Laboratory Manual for _Conceptual Integrated Science,_ © 2007 Addison Wesley

Summing Up

1. Current is passing through the center of a platform that supports four compasses. You are looking straight down at the platform. What is the direction of the current in each configuration shown below: coming toward you or going away from you?

a.

b.

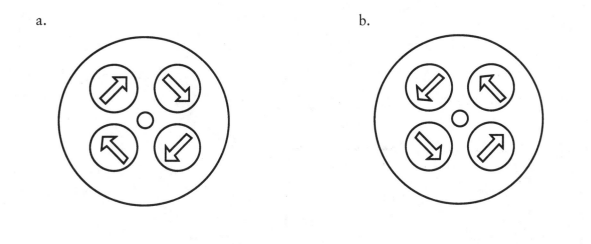

_____   _____

2. The direction of the magnetic field around a wire can be related to the direction of the current in the wire. If you imagine grabbing the wire with your thumb pointing in the direction of the current, your fingers wrap around the wire in the direction of the magnetic field. Which hand must you use for this exercise to give the correct relationship between the direction of the current and the direction of the magnetic field: your left or right?

_____

3. What is the source of **all** magnetic fields?

_____

Name _____ Date _____

Electricity and Magnetism: Magnetic Forces on Moving Charges

Motor Madness

Purpose
In this activity, you will investigate the principles that make electric motors possible.

Required Equipment and Supplies
handheld generator (Genecon® or equivalent)
connecting wires
about 50 centimeters (cm) of lead-free solder
2 collar hooks (or 2 10-cm lengths of lead-free solder)
about 30 cm of 1/4″ diameter wood dowel
support rod with base and rod clamp
2 bar magnets (strong alnico magnets are recommended)
small block of wood (about 2″ × 2″ × 1″)
2 rubber bands
2 D-cell batteries

Discussion
Perhaps the most important invention of the 19th century was the electric motor. You use a motor whenever you use electric power to make something move. A motor is used to start the engine of a car. Motors are used to spin compact discs. Motors are used to move elevators up and down. A list of motor applications would go on and on. But how do motors turn electric energy into mechanical energy? Let's find out!

An electric current can exert a force on a compass needle (which is simply a small magnet). But Newton's third law of motion suggests that something else must be going on here. What is it? Finish the statement:

   If an electric current can exert a force on a magnet, then a magnet

Procedure
Part A: The Magnetic Swing
Step 1: Arrange a solder "swing" by following the instructions below.

   a. Make a "sandwich" with the two bar magnets and the wood dowel as shown in Figure 1. The magnets must have opposite poles facing each other. Secure the sandwich with the rubber band. See Figure 1.

   b. Attach the support rod to the table clamp or ring stand base.

   c. Attach the wood dowel to the support rod.

   d. Place the collar hooks on the wood dowel about 10 cm apart. (You can use the short lengths of solder to make two hooks if collar hooks are not available.)

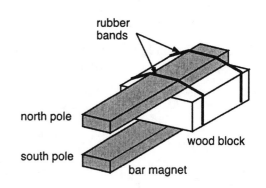

Figure 1

e. Bend the long length of solder into a square U-shaped "swing." Make wide hooks at the ends and hang the swing from the solder hooks on the wood dowel. **The swing must sway freely on its hooks.** See Figures 2 and 3.

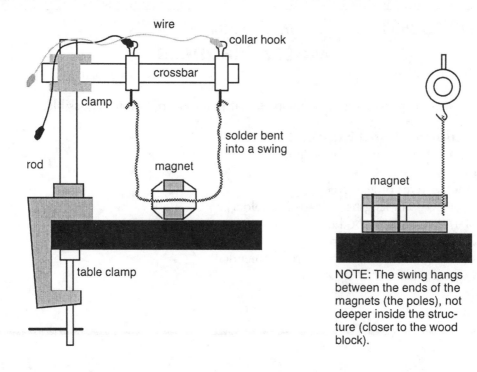

Figure 2. Front view of arrangement.   **Figure 3.** Side view of arrangement.

f. Arrange the height of the crossbar so that the bottom of the solder swing hangs between the magnets.

g. Attach the generator leads to the top leads of the hooks.

Step 2: Crank the generator one way. This will send electric current one way through the swing. Then crank the generator the other way.

1. What effect—if any—does the magnetic field of the bar magnets have on the current in the swing?

_____

2. How do you know that this effect is caused by the current's interaction with the bar magnets? (Would the swing sway if the magnets weren't there?)

_____

Step 3: Try "pumping" the swing by cranking the generator back and forth.

3. Do your observations of the magnetic swing confirm or contradict what you stated in the Discussion section above?

_____

Laboratory Manual for *Conceptual Integrated Science,* © 2007 Addison Wesley

Part B: The Simple Motor

Once it was found that a magnetic field could exert a force on an electric current, clever engineers designed practical ways to harness this force. They started to build electric motors. A motor transforms electric energy into mechanical energy. Some simple motors are made of coils of wire and magnets arranged so that when electric current flows through the wires, some part of the motor rotated.

The hand generator you've been using in this and other labs is, in fact, **a motor!**

Step 1: Hold the grip of the generator but not the crank handle. Touch the two leads of the generator to opposite terminals of a single D-cell battery. What happens?

_____

Step 2: Put two batteries together in series (in a line end to end) and touch the leads of the generator to opposite terminals of the arrangement. How is the result different from what happened in Step 1?

_____

Summing Up

1. The magnetic swing swayed due to interaction between the current in the wire and the magnetic field of the bar magnets. What are some ways this force could be made stronger (and thereby push the swing further in or further out)?

_____

_____

_____

2. Which of the devices listed below uses a motor?

___alarm clock	___toilet	___shower
___blow dryer	___shaver	___cassette player
___CD/DVD player	___radio	___vending machine
___lightbulb	___computer	___TV
___VCR	___washing machine	___car

3. List two more devices that use motors.

_____

_____

CONCEPTUAL INTEGRATED SCIENCE | Activity

Electricity and Magnetism: Electromagnetic Induction

Generator Activator

Purpose
In this activity, you will investigate electromagnetic induction, the principle behind electric generators.

Required Equipment and Supplies
bar magnet (strong alnico magnet is recommended)
air core solenoid or a long wire coiled into many loops
connecting wires
galvanometer ($-500\ \mu A - 0 - +500\ \mu A$ recommended)
hand-held generator (Genecon® or equivalent)
access to a second hand-held generator

Discussion
In 1820, Hans Christian Ørsted found that electricity could create magnetism. Scientists were convinced that if electricity could create magnetism, magnetism could create electricity. Still, 11 years would pass before the induction of electricity from magnetism would be discovered and understood. The best minds of the day set out to make this widely anticipated discovery, but it was Michael Faraday who put all the pieces together. In this activity, we will use a magnet to create an electric current and see how this effect is applied in electric generators.

Procedure
Part A: Electromagnetic Induction
Step 1: Connect the galvanometer to the coil (air core solenoid or looped wire).

Step 2: Determine a method for producing current in the coil using the bar magnet.

1. Describe your findings.

_____

_____

2. Can current be produced if

   a. the coil is at rest? If so, how?

   _____

   b. the magnet is at rest? If so, how?

   _____

   c. both the coil and the magnet are at rest? If so, how?

   _____

Part B: The Generator

Once it was found that a changing magnetic field could induce an electric current, clever engineers figured out practical ways to harness induction. They started to build electric generators. A generator transforms mechanical energy into electrical energy. Some simple generators are made of coils of wire and magnets arranged so that when some part of the generator is rotated, and electric current moves through the wires.

The hand-crank generator you have used in previous labs is such a device.

Step 1: Disconnect the coil from the galvanometer and attach the leads of the hand-crank generator to the galvanometer. Slowly turn the handle until the meter responds.

3. What does the galvanometer show?

_____

4. What happens if you turn the handle the other way?

_____

The invention of the generator allowed the production of continually flowing electrical energy without the use of chemical batteries. Further developments led to the wide-scale distribution of electrical energy and the availability of household electricity.

Large-scale electrical distribution grids are powered by large-scale generators. These generators are commonly powered by steam turbines. The heat used to generate the steam is typically produced by burning coal or oil, or as a byproduct of controlled nuclear reactions.

Step 3: Use a generator to power a motor! Connect two hand-crank generators to each other. Crank the handle of one of the generators.

5. What happens to the handle of the other generator? Would you say the other generator is acting as a motor?

_____

6. Not all the energy you put into the generator turns into mechanical energy in the motor. What is your evidence of this?

_____

Summing Up

1. Name each device described below

   a. Transforms electric energy into mechanical energy: _____

   b. Transforms chemical energy into electrical energy: _____

   c. Transforms mechanical energy into electrical energy: _____

2. What happens to the energy lost between the generator and motor in Step 3 (Part B) above?

_____

3. A classmate suggests that a generator could be used to power a motor that could then be used to power the generator. What do you think about this proposal and why?

_____

CONCEPTUAL INTEGRATED SCIENCE	Activity

Waves—Sound: Vibrations and Waves

Slow-Motion Wobbler

Purpose
In this activity, you will observe and explore the oscillation of a tuning fork.

Required Equipment and Supplies
low-frequency tuning forks [40–150-hertz (Hz) forks work best for large amplitudes]
strobe light, variable frequency

Discussion
The tines of a tuning fork oscillate at a precise frequency. That's why musicians use them to tune instruments. In this activity, you will investigate their motion with a special illumination system—a **stroboscope.**

Procedure
Strike a tuning fork with a mallet or the heel of your shoe (do **not** strike against the table or other hard object). Does it appear to vibrate? Try it again, this time dipping the tip of the tines just below the surface of water in a beaker. What do you observe?

_____

Now dim the room lights and strike a tuning fork while it is illuminated with a strobe light. For best effect, use the tuning fork with the longest tines available. Adjust the frequency of the strobe so that the tines of the tuning fork appear to be stationary. Then carefully adjust the strobe so that the tines slowly wag back and forth. Describe your observations.

_____

_____

Summing Up
1. What happens to the air next to the tines as they oscillate?

_____

2. Strike a tuning fork and observe how long it vibrates. Repeat, placing the handle against the tabletop or counter. Although the sound is louder, does the *time* the fork vibrates increase or decrease? Explain.

_____

_____

3. What would happen if you struck the tuning fork in outer space?

_____

CONCEPTUAL INTEGRATED SCIENCE | Activity

Waves—Sound: Interference

Sound Off

Purpose
In this demonstration, you will hear a dramatic effect of the interference of sound.

Required Equipment and Supplies
stereo radio, tape, or CD player with two movable speakers, one of which has a DPDT (double-pole double-throw) switch or a means of reversing polarity

Discussion
Interference is a behavior common to all waves. With water waves we see it in regions of calm where overlapping crests and troughs coincide. We see the effects of interference in the colors of soap bubbles and other thin films where reflection from nearby surfaces puts crests coinciding with troughs. In this activity, we'll dramatically experience the effects of interference with sound!

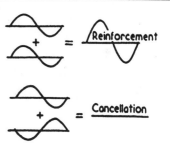

Procedure
Play the stereo player with both speakers in phase (with the plus and minus connections to each speaker the same). Play it in mono mode so the signals of each speaker are identical. Note the fullness of the sound. Now reverse the polarity of one of the speakers (either by physically interchanging the wires or by means of the switch provided). Note the sound is different—it lacks fullness. Some of the waves from one speaker are arriving at your ear out of phase with waves from the other speaker.

Now place the speakers facing each other at arm's length. The long waves are interfering destructively, detracting from the fullness of the sound. Gradually bring the speakers closer to each other. What happens to the volume and fullness of the sound heard? Bring them face to face against each. What happens to the volume now?

Summing Up
1. What happens to the volume of sound when the face-to-face speakers are switched so both are in phase?

_____

2. Why is the volume so diminished when the out-of-phase speakers are brought together face to face? And why is the remaining sound so "tinny"?

_____

3. What practical applications can you think of for canceling sound?

_____

CONCEPTUAL INTEGRATED SCIENCE	Activity

Waves—Light: The Nature of Light

Pinhole Image

Purpose
To investigate the operation of a pinhole "lens" and compare it to the eye.

Required Equipment and Supplies
3" × 5" card
straight pin
meterstick

Discussion
The image cast through a pinhole in a pinhole camera has the property of being in clear focus at any distance from the pinhole. That's because the tinyness of the pinhole does not allow the overlapping of light rays. (The tinyness also doesn't allow for the passage of much light, so pinhole images are normally dim as well.) When a pinhole is placed at the center of the pupil of your eye, the light that passes through the pinhole forms a focused image no matter where the object is located. Pinhole vision, although dim, is remarkably clear. In this activity, you will use a pinhole to see fine detail more clearly.

Procedure
Step 1: Bring this printed page closer and closer to your eye until you cannot clearly focus on it any longer. Even though your pupil is small, your eye does not act like a true pinhole camera because it does not focus well on nearby objects.

Step 2: Poke a single pinhole into a card. Hold the card in front of your eye and read these instructions through the pinhole. Bright light on the print may be required. Bring the page closer and closer to your eye until it is a few centimeters away. You should be able to read the type clearly. Then quickly remove the card and see if you can still read the instructions without the benefit of the pinhole.

Summing Up
Enlist the help of people in your lab who are nearsighted and who are farsighted (if you're not one of them yourself).

1. A farsighted person without corrective lenses cannot see close-up objects clearly. Can a farsighted person without corrective lenses see close-up objects clearly through a pinhole?

_____

2. A nearsighted person without corrective lenses cannot see far-away objects clearly. Can a nearsighted person without corrective lenses see far-away objects clearly through a pinhole?

_____

3. Why does a page of print appear sharper yet dimmer when seen through the pinhole?

_____

CONCEPTUAL INTEGRATED SCIENCE	Activity

Waves—Light: The Nature of Light

Pinhole Camera

Purpose
In this activity, you will observe images formed by a simple convex lens and compare cameras with and without a lens.

Required Equipment and Supplies
covered shoebox
25-millimeter (mm) converging lens
tracing paper
aluminum foil
masking tape

Discussion
The first camera used a pinhole opening to let light in. Because the hole is so small, light rays that enter cannot overlap. This is why a clear image is formed on the inner back wall of the camera. Because the opening was small, a long time was required to expose the film sufficiently. A lens allows more light to pass through and still focus the light onto the film. Cameras with lenses require much less time for exposure, and the pictures came to be called "snapshots."

Procedure
Step 1: Construct a camera as shown in Figure 1. It is a shoebox with a hole about an inch or so in diameter on one end, some glassine paper taped in the center to act as a screen, and an opening for viewing the screen on the other end. Tape some foil over the lens hole of the box. Poke a pinhole in the middle of the foil. Point the camera toward a brightly illuminated scene, such as the window during the daytime. Light enters the pinhole and falls on the glassine paper. Observe the image of the scene on the glassine paper.

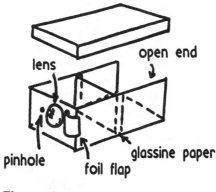

Figure 1

1. Is the image on the screen upside down (inverted)

_____

2. Is the image on the screen reversed left to right?

_____

Step 2: Now remove the pinhole foil and tape a lens over the hole in the box. You now have a lens camera. Move it around and watch people or other scenery.

3. Is the image on the screen upside down (inverted)?

_____

4. Is the image on the screen reversed left to right?

_____

Step 3: Unlike a lens camera, pinhole cameras focus equally well on objects at practically all distances. Aim the camera lens at nearby objects and see if the lens focuses them.

5. Does the lens focus on nearby objects as well as it does on distant ones?

_____

Step 4: Draw a ray diagram as follows. First, draw a ray for light that passes from the top of a distant object through a pinhole and onto a screen. Second, draw another ray for light that passes from the bottom of the object through the pinhole and onto the screen. Then sketch the image created on the screen by the pinhole.

Summing Up

1. Why is the image created by the pinhole dimmer than the one created by the lens?

_____

_____

2. How is a pinhole camera similar to your eye? Do you think that the images formed on the retina of your eye are upside down? Your explanation might include a diagram.

_____

_____

_____

Name _____ Date _____

Waves—Light: Reflection

Mirror, Mirror, on the Wall . . .

Purpose
In this activity, you will investigate the minimum size mirror required for you to see a full image of yourself.

Required Equipment and Supplies
large mirror, preferably full length
ruler and masking tape

Discussion
Why do shoe stores and clothier shops have full-length mirrors? Need a mirror be as tall and wide as you for you to see a complete image of yourself?

Procedure
Step 1: Stand about arm's length in front of a vertical full-length mirror. Reach out and place a small piece of masking tape on the image of the top of your head. Now stare at your toes. Place the other piece of tape on the mirror where your toes are seen. Use a meter stick to measure the distance from top of your head to your toes. How does the distance between the pieces of tape on the mirror compare to your height?

_____

Step 2: Now stand about 3 meters from the mirror and repeat. Stare at the top of your head and toes and have an assistant move the tape so that the pieces of tape mark where head and feet are seen. Move further away or closer, and repeat. What do you discover?

_____

Summing Up
1. Does the location of the tape depend on your distance from the mirror?

_____

2. What is the shortest mirror you can use to see your entire image? Do you **believe** it?

_____

Going Further
Try this one if a full-length mirror is not readily available *or* you are a disbeliever! Hold a ruler next to your eye. Measure the height of a common pocket mirror. Hold the mirror in front of you so that the image includes the ruler. How many centimeters of the ruler appear in the image? How does this compare to the height of the mirror?

_____

CONCEPTUAL INTEGRATED SCIENCE	Experiment

The Atom: Atoms are Ancient, Tiny, and Empty

Thickness of a BB Pancake

Purpose
In this experiment, you will determine the diameter of a BB without actually measuring it. (The abbreviation "BB" stands for "ball bearing.")

Required Equipment and Supplies
75 milliliters (mL) of BB shot
100-mL graduated cylinder
tray
ruler
micrometer

Discussion
This activity distinguishes between *area* and *volume* and sets the stage for the follow-up experiment "Oleic Acid Pancake," where you will estimate the size of molecules. To see the difference between area and volume, consider eight wooden blocks arranged to form a single 2 × 2 × 2-inch cube. Because any cube has six sides, the outer surface area will be six times the area of one face—6 × 4 = 24 square inches (in²). The cube form exposes a minimum surface area (which is why buildings in cold areas are most often cubical in shape). But what if the blocks were put together differently? For example, consider an arrangement where all the blocks are placed together within a single layer. Do you see that the total surface area would be greater than 28 in²? Discuss this with your lab partners.

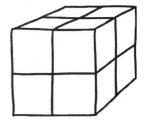

The different configurations have different surface areas, but the volume remains constant. The volume of pancake batter is also the same whether it is in the mixing bowl or spread out on a surface. The volume of a pancake equals the surface area of one flat side multiplied by the thickness. If both the volume and the surface area are known, then the thickness can be calculated.

$$\text{Volume} = \text{area} \times \text{thickness}$$

so simple rearrangement gives

$$\text{Thickness} = \frac{\text{volume}}{\text{area}}$$

Instead of cubical blocks or pancake batter, consider a graduated cylinder that contains BBs. The space taken up by the BBs is easily read as volume on the side of the cylinder. If the BBs are poured into a tray, their volume remains the same. Can you think of a way to estimate the diameter (or thickness) of a single BB without measuring the BB itself? Try it in this activity and see. In the next lab, you will use this same technique to estimate the size of a molecule.

Procedure

Step 1: Use a graduated cylinder to measure the volume of the BBs. (Note that 1 mL = 1 cm³.)

Volume = _____ cm³

Step 2: Carefully spread the BBs out to make a compact layer one pellet thick on the tray. With a ruler, determine the area covered by the BBs. Describe your procedure and show your computations.

Area = _____ cm²

Step 3: Using the area and volume of the BBs, estimate the thickness (diameter) of a BB. Show your computations.

Estimated thickness = _____ cm

Step 4: Check your estimate by using a micrometer to measure the thickness (diameter) of a BB.

Measured thickness = _____ cm

Summing Up

1. How does your estimate compare to the measurement of the diameter of the BB? Calculate the percentage error (consult the Appendix on how to do this) between the estimated and measured thickness of the BB.

_____

_____

2. Oleic acid is an organic substance that is soluble in alcohol but insoluble in water. When a drop of oleic acid is placed in water, it usually spreads out over the water surface to create a *monolayer,* a layer that is one molecule thick. From your experience with BBs, describe a method for estimating the size of an oleic acid molecule.

_____

_____

_____

Name _____ Date _____

| CONCEPTUAL INTEGRATED SCIENCE | **Experiment** |

Atoms are Ancient, Tiny, and Empty
Oleic Acid Pancake

Purpose
In this experiment, you will estimate the size of a single molecule of oleic acid.

Required Equipment and Supplies
tray
water
chalk dust or lycopodium powder
eyedropper
oleic acid solution (5 milliliters (mL) oleic acid in 995 mL of ethanol)
10-mL graduated cylinder

Discussion
During this experiment, you will estimate the diameter of a single molecule of oleic acid! The procedure for measuring the *diameter* of a molecule will be much the same as that of measuring the diameter of a BB in the previous activity. The diameter is calculated by dividing the volume of the drop of oleic acid by the area of the *monolayer* film that is formed. The diameter of the molecule is the depth of the monolayer.

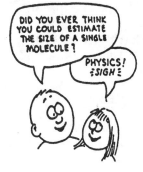

$$\text{Volume} = \text{area} \times \text{depth}$$

$$\text{Depth} = \frac{\text{volume}}{\text{area}}$$

Procedure
Step 1: Pour water into the tray to a depth of about 1 cm. So that the acid film will show itself, spread chalk dust or lycopodium powder very lightly over the surface of the water; too much will hem in the oleic acid.

Step 2: Using the eyedropper, gently add a single drop of the oleic acid solution to the surface of the water. When the drop touches the water, the alcohol in it will dissolve in the water, but the oleic acid will not. The oleic acid spreads out to form a nearly circular patch on the water. Measure the diameter of the oleic acid patch in several places, and compute the average diameter of the circular patch.

Average diameter = _____ cm

The average radius is, of course, half the average diameter. Now compute the area of the circle ($A = \pi r^2$).

Area of circle = _____ cm^2

Step 3: Count the number of drops of solution needed to occupy 1 mL (or 1 cm^3) in the graduated cylinder. Do this three times, and find the average number of drops in 1 cm^3 of solution.

Number of drops in 1 cm^3 = _____

Divide 1 cm³ by the number of drops in 1 cm³ to determine the volume of a single drop.

Volume of single drop = _____ cm³

Step 4: The volume of the oleic acid alone in the circular film is much less than the volume of a single drop of the solution. The concentration of oleic acid in the solution is 5 mL per liter of solution. Every cubic centimeter of the solution thus contains only $\frac{5}{1000}$ cm³, or 0.005 cm³, of oleic acid. The volume of oleic acid in one drop is thus 0.005 of the volume of one drop. Multiply the volume of a drop by 0.005 to find the volume of oleic acid in the drop. This is the volume of the layer of acid in the tray.

Volume of oleic acid = _____ cm³

Step 5: Estimate the diameter of an oleic acid molecule by dividing the volume of oleic acid by the area of the circle.

Diameter = _____ cm

The diameter of an oleic acid molecule as obtained by this method is good, but not precise. This is because an oleic acid molecule is not spherical, but rather elongated like a hot dog. One end is attracted to water, and the other end points away from the water surface. The molecules stand up like people in a puddle! So the estimated diameter is actually the estimated length of the short side of an oleic acid molecule.

Summing Up

1. What is meant by a *monolayer?*

_____

2. Why is it necessary to dilute the oleic acid for this experiment? Why alcohol?

_____

3. The shape of oleic acid molecules is more like that of a hot dog than a sphere. Furthermore, one end is attracted to water (*hydrophilic*) so that the molecule stands up on the surface of water. Assume an oleic molecule is 10 times longer than it is wide. Then estimate the volume of one oleic acid molecule.

_____

_____

Laboratory Manual for *Conceptual Integrated Science,* © 2007 Addison Wesley

CONCEPTUAL INTEGRATED SCIENCE	Experiment

The Atom: Atomic Spectra

Bright Lights

Purpose

In this activity, you will examine light emitted by various elements when heated and identify a component in an unknown salt by examining the light emitted by the heated salt.

Required Equipment and Supplies

Equipment
Bunsen burner
heatproof glove or pot holder
metal spatulas or flame test wires
diffraction gratings
spectroscope (commercial or homemade)
ring stand with clamp large enough to
    fit around spectroscope barrel
colored pencils
discharge tubes containing various gases

Chemicals
0.1 M HCl solution
lithium chloride powder
sodium chloride powder
potassium chloride powder
calcium chloride powder
strontium chloride powder
barium chloride powder
cupric chloride powder

Discussion

When materials are heated, their elements emit light. The color of the light is characteristic of the type of elements in the heated material. Lithium, for example, emits red light, and copper emits green light. An element emits light when its electrons make a transition from a higher energy level to a lower energy level. Every element has its own characteristic pattern of energy levels and therefore emits its own characteristic pattern of light frequencies (colors) when heated.

It is interesting to look at the light emitted by elements through either a diffraction grating or a prism. Rather than produce a continuous spectrum of colors, the elements produce a spectrum that is discontinuous, showing only particular colors (frequencies). When the emitted light passes first through a thin slit, and then through the grating or prism, the different colors appear as a series of vertical lines, as shown in Figure 1. Each vertical line corresponds to a particular energy transition for an electron in an atom of the heated element. The pattern of lines, referred to as an emission spectrum, is characteristic of the element and is often used as an identifying feature—much like a fingerprint. Astronomers, for instance, can tell the elemental composition of stars by examining their emission spectra.

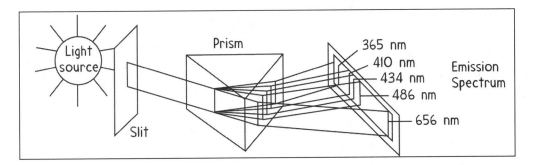

Figure 1. When heated, a gaseous element produces a discontinuous emission spectrum.

Procedure

Part A: Flame Tests

Step 1: Using either a heatproof glove or a pot holder, hold the tip of a metal spatula in a Bunsen burner flame until the spatula tip is red hot, and then dip it in a 0.1 *M* HCl solution. Repeat this cleaning process several times until you no longer see color coming from the metal when it is heated.

Step 2: Obtain small amounts of the metal salts to be tested, and label each sample. Dip the spatula tip first into the HCl solution and then into one of the salts, so that the tip becomes coated with the powder. Then put the tip into the flame and observe the color. Record your observations on the report sheet.

Step 3: Rinse the spatula in water, and then clean as described in Step 1.

Step 4: Repeat this procedure for all the salts, being sure to clean the spatula each time.

Part B: Flame Tests Using a Spectroscope

Step 1: Use the same procedure as in Part A, this time observing the flame through a spectroscope mounted on a ring stand. You can use either a commercial unit or a homemade one like the one shown in Figure 2.

Step 2: On your report sheet, sketch the predominant lines you observe for each salt, using colored pencils. You will see lines both to the left and to the right of the slit. Sketch only the lines to the right. (Note: some salts will also show regions of continuous color.)

Step 3: Obtain an unknown metal salt from your instructor and record its number.

Step 4: Observe and sketch the line spectrum for your unknown, using colored pencils.

Step 5: Identify your unknown based on its spectrum.

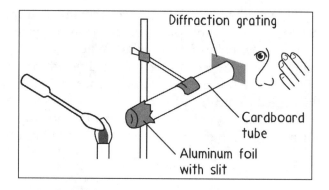

Figure 2. A homemade spectroscope.

Part C: Gas Discharge Tubes

Step 1: Observe through a diffraction grating the light emitted from various gas discharge tubes. (You need not pass the light through a slit because the discharge tubes are narrow.)

Step 2: Using colored pencils, sketch the line spectra for all samples available, especially hydrogen, oxygen, and water vapor.

Bright Lights Report Sheet

Part A: Flame Test

Compound	LiCl	NaCl	KCl	CaCl$_2$	SrCl$_2$	BaCl$_2$	CuCl$_2$
Color							

Part B: Flame Tests Using a Spectroscope

Unknown number _____

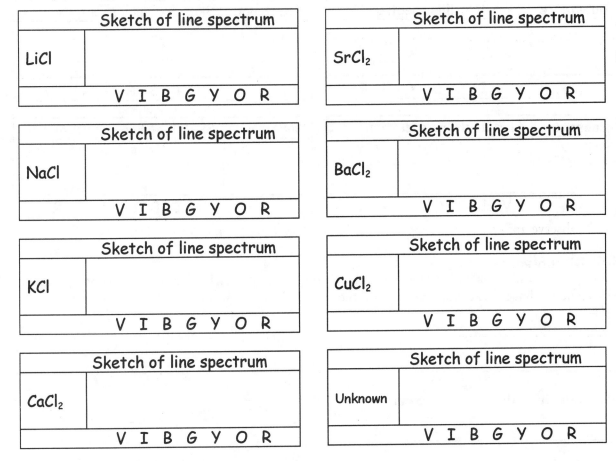

	Sketch of line spectrum
LiCl	
	V I B G Y O R

	Sketch of line spectrum
SrCl$_2$	
	V I B G Y O R

	Sketch of line spectrum
NaCl	
	V I B G Y O R

	Sketch of line spectrum
BaCl$_2$	
	V I B G Y O R

	Sketch of line spectrum
KCl	
	V I B G Y O R

	Sketch of line spectrum
CuCl$_2$	
	V I B G Y O R

	Sketch of line spectrum
CaCl$_2$	
	V I B G Y O R

	Sketch of line spectrum
Unknown	
	V I B G Y O R

Identity of unknown: _____

Part C: Discharge Tubes

Substance	Sketch of line spectrum
V I B G Y O R	

Substance	Sketch of line spectrum
V I B G Y O R	

Substance	Sketch of line spectrum
V I B G Y O R	

Substance	Sketch of line spectrum
V I B G Y O R	

Substance	Sketch of line spectrum
V I B G Y O R	

Substance	Sketch of line spectrum
V I B G Y O R	

Substance	Sketch of line spectrum
V I B G Y O R	

Substance	Sketch of line spectrum
V I B G Y O R	

Summing Up

1. When the spatula was initially being cleaned in the flame, it may have given off yellow light. If this happened, what residue was probably on the spatula before it was cleaned?

_____

_____

2. What produces the colors of fireworks?

_____

_____

3. Is the gas in a blue "neon lamp" actually neon? Explain.

_____

_____

4. Does the line spectrum of water vapor bear any resemblance to the line spectra of hydrogen and oxygen? Why or why not?

Laboratory Manual for *Conceptual Integrated Science,* © 2007 Addison Wesley

CONCEPTUAL INTEGRATED SCIENCE	Activity

Nuclear Physics: Radioactivity

Get a Half-Life!

Purpose
In this activity, you will simulate radioactive decay half-life.

Required Equipment and Supplies
25 small color-marked cubes per group (one side red, two sides blue, three sides blank). (Spray-painted sugar cubes work well. Multifaceted dice may also be used.)

Discussion
The rate of decay for a radioactive isotope is measured in terms of **half-life**—the time for one half of a radioactive quantity to decay. Each radioactive isotope has its own characteristic half-life (Table 1). For example, the naturally occurring isotope of uranium, uranium-238, decays into thorium-234 with a half-life of 4,510,000,000 years. This means that only half of an original amount of uranium-238 remains after this time. After another 4,510,000,000 years, half of this decays leaving only one-fourth of the original amount remaining. Compare this with the decay of polonium-214, which has a half-life of 0.00016 seconds. With such a short half-life, any sample of polonium-214 will quickly disintegrate.

Table 1

Isotope	Half-life
Uranium-238	4,510,000,000 years
Plutonium-239	24,400 years
Carbon-14	5,730 years
Lead-210	20.4 years
Bismuth-210	5.0 days
Polonium-214	0.00016 seconds

The half-life of an isotope can be calculated by the amount of radiation coming from a known quantity. In general, the shorter the half-life of a substance, the faster it decays, and the more radioactivity per amount is detected.

In this activity, you will investigate three hypothetical substances, each represented by a color on the face of a cube. The first substance, represented by a given color, is marked on only one side of the cube. The second substance, represented by a second color, is marked on two sides of the cube, and the third substance, represented by a third color (or lack thereof), is marked on the remaining three sides. Rolling a large number of these identically painted cubes simulates the process of decay for these substances. As a substance's color turns face up, it is considered to have decayed and is removed from the pile. This process is repeated until all of the cubes have been removed. Because the color of the first substance is only on one side, this substance will decay the slowest (because its color will fall face up least frequently and it will stay in the game longer). The second substance, marked on two sides, will decay faster, requiring fewer rolls before all the cubes are removed. The third substance, marked on three sides, will decay the fastest. After tabulating and graphing the numbers of cubes that decay in each roll for these simulated substances, you will be able to determine their half-lives.

Procedure

Step 1: Shake the cubes in a container and roll them onto a flat surface.

Step 2: Count the one-side color faces that are up and record this number under "Removed" in the Data Table.

Step 3: Remove the one-side color cubes in a pile off to the side.

Step 4: Gather the remaining cubes back into the container and roll them again.

Step 5: Repeat Steps 2–4 until all cubes have been counted, tabulated, and set aside.

Step 6: Repeat Steps 1–5 removing cubes that show the two-side color faces up.

Data Table

Throw	First Substance (One-side color)		Second Substance (Two-side color)		Third Substance (Three-side color)	
	Removed	Remaining	Removed	Remaining	Removed	Remaining
Initial Count						
1						
2						
3						
4						
5						
6						
7						
8						
9						
10						
11						
12						
13						
14						
15						
16						
17						
18						
19						
20						

Laboratory Manual for *Conceptual Integrated Science,* © 2007 Addison Wesley

	First Substance (One-side color)		Second Substance (Two-side color)		Third Substance (Three-side color)	
Throw	Removed	Remaining	Removed	Remaining	Removed	Remaining
21						
22						
23						
24						
25						

Step 7: Repeat Steps 1–5 removing cubes that show the three-side color faces up.

Step 8: Plot the number of cubes remaining versus the number of throws for each substance on the following graph. Use a different color or line pattern to graph the results for each substance. For each substance, draw a single smooth line or curve that approximately connects all points. **Do not connect the dots!** Indicate your color or line pattern code below the graph.

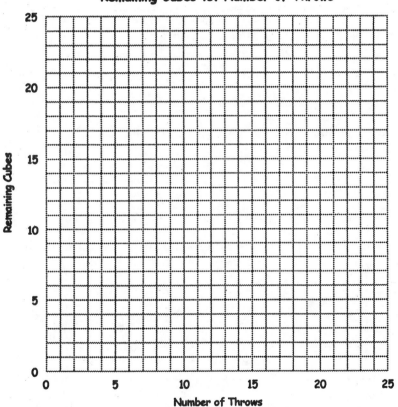

First substance (one side on cube) color or line pattern:

Second substance (two sides on cube) color or line pattern:

Third substance (three sides on cube) color or line pattern:

Summing Up

1. How many rolls did it take for the number of each colored cube to be reduced by half? These are your half-life readings.

   One-side color _____   Two-side color _____   Three-side color _____

2. The half-life of a decaying substance is measured in units of time. What is the unit of half-life used in this simulation?

   _____

3. In each case, how many rolls did it take to remove all of the cubes?

   One-side color _____   Two-side color _____   Three-side color _____

4. Which of these hypothetical substances would be the most harmful?

   _____

5. How might you simulate the radioactive decay of a substance that decays into a second substance that also decays?

   _____

   _____

6. Is it possible to estimate the half-life of a substance in a single throw? How accurate might this estimate be?

   _____

   _____

7. Are your lines in the graph for Step 8 fairly straight or do they curve? Do these lines correspond to a constant or nonconstant rate of decay?

   _____

   _____

8. a. Substance X has a half-life of 10 years. If you start with 1000 g, how much will be left after:

   i.   10 years? _____

   ii. 20 years? _____

   iii. 50 years? _____

   iv. 100 years? _____

   b. Will this sample of substance X ever totally disappear? If so, estimate how soon. If not, explain.

   _____

   _____

Laboratory Manual for *Conceptual Integrated Science,* © 2007 Addison Wesley

CONCEPTUAL INTEGRATED SCIENCE	Activity

Nuclear Physics: Nuclear Fission

Chain Reaction

Purpose
In this activity, you will simulate a simple chain reaction.

Required Equipment and Supplies
100 dominoes
large table or floor space
stopwatch

Discussion
Give your cold to two people who, in turn, give it to two others who, in turn, do the same on down the line, and before you know it, everyone in class is sneezing. You have set off a chain reaction. Similarly, when one electron in a photomultiplier tube in certain electronic instruments hits a target that releases two electrons that, in turn, do the same on down the line, a tiny input produces a large output. When one neutron triggers the release of two or more neutrons in a piece of uranium, and the triggered neutrons trigger others in succession, the results can be devastating. In this activity, we'll explore this idea.

Procedure
Step 1: Set up a strand of dominoes about half a domino length apart in a straight line. Gently push the first domino over, and measure how long it takes for the entire strand to fall over (just like in those television commercials).

Step 2: Arrange the dominoes as in Figure 1, so that when one domino falls, another one or two are toppled over. These topple others in chain reaction fashion. Set up until you run out of dominoes or table space. When you finish, push the first domino over and watch the reaction. Notice the number of the falling dominoes per second at the beginning versus the end.

Figure 1

Summing Up
1. Which reaction, wide spaced or close spaced dominoes, took a shorter time?

_____

_____

2. How did the number of dominoes being knocked over per second change for each reaction?

_____

_____

3. What caused each reaction to stop?

_____

4. Now imagine that the dominoes are the neutrons released by uranium atoms when they fission (split apart). Neutrons from the nucleus of a fissioning uranium atom hit other uranium atoms and cause them to fission. This reaction continues to grow if there are no controls. Such an uncontrolled reaction occurs in a split second and is called a *nuclear explosion*. How is the domino chain reaction similar to the nuclear fission process?

_____

_____

_____

_____

5. How is the domino reaction dissimilar to the nuclear fission process?

_____

_____

_____

_____

Laboratory Manual for *Conceptual Integrated Science,* © 2007 Addison Wesley

Name _____ Date _____

Investigating Matter: Physical and Chemical Properties
Chemical Personalities

Purpose
In this activity, you will identify various physical and chemical properties of matter and distinguish between physical and chemical changes.

Required Equipment and Supplies

Equipment	Chemicals
thermometer with cork holder	various elements and compounds
ring stand with two clamps	methanol
hot plate	iodine crystals
250-milliliter (mL) and 500-mL beakers	sucrose crystals
large test tube	acetone
boiling chips	steel wool
glass stirring rod	cupric sulfate pentahydrate crystals
well-plates (ceramic or glass)	10% sodium carbonate solution
eyedroppers	10% sodium sulfate solution
microspatula	3 M HCl
evaporating dishes	10% sodium chloride solution
ice	10% calcium chloride solution
medium test tube with stopper	luminol crystals
	sodium perborate crystals

Discussion
Chemistry is the study of matter. It is very common for a chemist to need to describe a bit of matter as thoroughly as possible. In doing so, the chemist would certainly list *physical properties*. Many physical properties can be observed using our senses; color, crystal shape, and phase at room temperature are some examples. Other physical properties involve quantitative observations and so must be measured; density, specific heat, and boiling point are three examples. A physical change is any change in a substance that does not involve a change in its chemical composition. During a physical change, no new chemical bonds are formed, and so the chemical composition remains the same. Examples of physical change are boiling, freezing, expanding, and dissolving.

Matter can also be characterized by its chemical properties. The *chemical properties* of a substance include all the chemical changes possible for that substance. A chemical change is one in which the substance is transformed to a new substance. That is, there is a change in the chemical composition of the substance. During a chemical change, the atoms are pulled apart from one another, rearranged, and put back in a new arrangement. Examples of chemical change are burning, rusting, fermenting, and decomposing.

In this experiment, you will first identify and record various physical properties of substances, using qualitative observations, such as changes in color or phase, and quantitative observations, such as boiling points. In the second part, you will look at changes in matter and determine if they are physical or chemical.

Procedure

Part A: Physical Properties

Step 1: Examine the various substances provided by your instructor and record your observations in Table 1 of the report sheet. (Note: some substances may be toxic. As a precaution, do not open any containers without the permission of your instructor.)

Step 2: Assemble the apparatus shown in Figure 1. Add about 300 mL of water to the 500-mL beaker and about 3 mL of methanol, CH_3OH, to the test tube. Do not forget to add the boiling chips to the test tube. Turn on the hot plate to medium. Use a stirring rod to stir the water while it is being heated, and pay close attention to the thermometer readings. Note that as the methanol boils, its temperature remains constant. Record the boiling point.

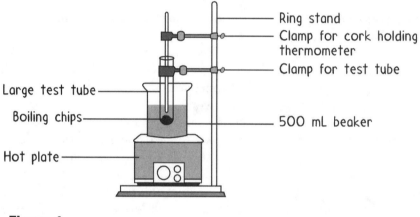

Figure 1

Step 3a: Place a small crystal of iodine in a well of one well-plate and a small crystal of sucrose in a well of a second well-plate. Use an eyedropper to fill each well with distilled water and stir gently with a microspatula. Record whether each substance is completely soluble, partially soluble, or insoluble. Rinse the iodine into a designated waste container and the sucrose into the sink.

Step 3b: Repeat the procedure using acetone as the solvent. You may need to rinse the iodine into another designated waste container (ask your instructor). The sucrose can be rinsed into the sink with water.

Laboratory Manual for *Conceptual Integrated Science,* © 2007 Addison Wesley

Part B: Physical and Chemical Changes

Complete Table 2 of the report sheet for each of the following systems.

Step 1: Inspect a small piece of steel wool. Place it in an evaporating dish, and heat on a hot plate set to high. Allow the system to cool to room temperature. Observe and record any changes in the steel wool.

Step 2: Inspect some cupric sulfate pentahydrate crystals, $CuSO_4 \cdot 5H_2O$. Place a few crystals in an evaporating dish, and heat on a hot plate set to medium. Observe and record any changes in the salt. After the system has cooled to room temperature, add a few drops of water to the crystals. Observe and record any changes.

Step 3: Place a few drops of a 10% sodium carbonate solution, Na_2CO_3, in one well of a well-plate and a few drops of a 10% sodium sulfate solution, Na_2SO_4, in a second well of the same well-plate. Add two or three drops of 3 M hydrochloric acid to each well. Observe and record any changes.

Step 4: Place a few drops of a 10% sodium chloride solution, NaCl, in one well of a well-plate and a few drops of a 10% calcium chloride solution, $CaCl_2$, into a second well of the same well-plate. Add several drops of a 10% sodium carbonate solution to each well. Observe and record any changes.

Step 5: Inspect some iodine crystals, I_2. Place a few of the crystals in a dry 250-mL beaker and cover with an evaporating dish that contains ice, as shown in Figure 2. In a fume hood, place the beaker on a hot plate set to medium. Observe and record any changes.

Step 6: Fill a medium test tube that can be stoppered about halfway with distilled water. Using a microspatula, add one scoop of luminol crystals and one scoop of sodium perborate crystals. Then add one small crystal of cupric sulfate pentahydrate. Quickly stopper the test tube, and take it to a darkened room. Observe and record any changes.

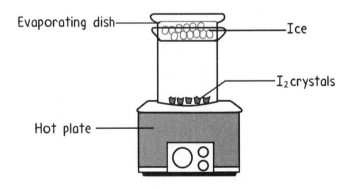

Figure 2

Physical and Chemical Properties and Changes Report Sheet

Part A: Physical Properties

1. Complete Table 1.

Table 1

Name of Substance	Chemical Formula	Phase at Room Temperature	Color	Other Physical Properties Observed	Element or Compound?

2. Boiling point of methanol _____ °C

3. Solubility:

Iodine in water _____   Sucrose in water _____

Iodine in acetone _____   Sucrose in acetone _____

Part B: Physical and Chemical Changes

1. Complete Table 2.

Table 2

Procedure	Observation	Physical Change or Chemical Change?	Evidence or Reasoning
1. steel wool + heat			
2a. $CuSO_4 \cdot 5H_2O$ + heat			
2b. $CuSO_4 + H_2O$			
3a. $Na_2CO_3 + HCl$			
3b. $Na_2SO_4 + HCl$			
4a. $NaCl + Na_2CO_3$			
4b. $CaCl_2 + Na_2CO_3$			
5. I_2 + heat			
6. Luminol + sodium perborate + $CuSO_4$			

Summing Up

1. Distinguish between a qualitative observation and a quantitative one. Give an example of each from this experiment.

2. Classify the following properties of sodium metal as physical or chemical:

   a. silver metallic color   _____

   b. turns gray in air   _____

   c. melts at 98°C   _____

   d. reacts explosively with chlorine   _____

3. Classify the following changes as physical or chemical:

   a. steam condenses to liquid water on a cool surface _____

   b. baking soda dissolves in vinegar, producing bubbles   _____

   c. mothballs gradually disappear at room temperature   _____

   d. baking soda loses mass as it is heated   _____

CONCEPTUAL INTEGRATED SCIENCE | Activity

The Nature of Chemical Bonds: Electron Shells and Chemical Bonding

Dot to Dot: Electron-Dot Diagrams

Purpose

To practice writing plausible electron-dot structures for simple molecules

Required Equipment and Supplies

periodic table

Discussion

To help predict how atoms bond together, you can use the *octet rule,* which states that atoms form chemical bonds so as to have a filled outermost occupied shell. Looking at a periodic table, we see that the noble gases already have filled the outermost occupied shells, which explains why they tend not to form chemical bonds. Note that all the noble gases except helium have *eight* outermost electrons, and hence, the name "octet rule."

The octet rule can be used to build an ammonia molecule, NH_3. Knowing that nitrogen has five valence electrons and a hydrogen one, you can satisfy the octet rule with the electron-dot structure

$$\text{H} \overset{\bullet\bullet}{\underset{\overset{\bullet\bullet}{\text{H}}}{\text{N}}} \text{H} \qquad \text{(eight electrons surrounding N, two electrons surrounding H)}$$

In this structure, the nitrogen has eight valence electrons so that its outermost occupied shell, which has a capacity for eight electrons, is filled. Each hydrogen has two valence electrons so that its outermost occupied shell, which has a capacity for two electrons, is also filled.

Using the octet rule, you will learn to draw electron-dot structures of simple covalent compounds given their chemical formulas. (This activity focuses on bonding in simple covalent molecules that obey the octet rule. There are many molecules that do not follow this rule, but they are not introduced in this laboratory.)

To construct a plausible electron-dot structure for a molecule,

1. Determine the total number of valence electrons available from the chemical formula by adding up the valence electrons of all atoms in the molecule.

2. Write the chemical symbols for all atoms in the molecule, arranging the symbols as you think they might appear in the molecule. Many molecules contain a central atom. If there are many elements present, place them in the order in which they are written in the formula.

3. Place single bonds between all pairs of atoms, remembering that each bond represents *two* electrons, one from each atom.

4. Add the appropriate number of electrons around each atom so that the atom obeys the octet rule (eight electrons around all atoms other than hydrogen, two around hydrogen).

5. Count the number of electrons used in your structure. If the number if equal to the sum you calculated in Step 1, your structure is plausible and you are done. If the number of electrons used is greater than the sum you calculated in Step 1, try putting in multiple bonds (try double first, then triple) until all atoms obey the octet rule and the number of electrons used matches the number calculated in Step 1.

6. If the number of electrons used is less than the sum you calculated in Step 2, you probably made an error in your count in Step 1 and so should recalculate.

Procedure

Part A: Drawing Electron-Dot Structures

To show these steps in action, here are examples using carbon tetrabromide and carbon monoxide.

1. Total number of valence electrons available:

Carbon tetrabromide (CBr_4)			Carbon monoxide (CO)		
1 C atom (4 valence electrons each)	$1 \times 4 =$	4	1 C atom (4 valence electrons)	$1 \times 4 =$	4
4 Br atoms (7 valence electrons each)	$4 \times 7 =$	28	1 O atom (6 valence electrons)	$1 \times 6 =$	6
Total		32	Total		10

2. Arrangement of symbols for all atoms (locate central atom if applicable):

```
        Br

   Br   C   Br              C   O

        Br
```

3. Single bonds between all pairs of atoms.

```
        Br
        ··
   Br : C : Br              C : O
        ··
        Br
```

4. Add remaining electrons to complete octets:

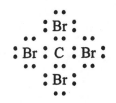

5. Count the number of electrons used:

The above structure uses 32 electrons, which is equal to the number calculated in Step 1. The structure is complete.

The above structure uses 14 electrons, more than the total number calculated in Step 1. A multiple bond must be present. With a double bond, the structure would be

Still, there is a problem because the structure uses 12 electrons. A triple bond must be present:

:C ⫶⫶ O:

Finally, the structure obeys the octet rule and uses the total number of valence electrons available. The structure is complete.

Many chemists prefer to use a line to represent a bonded pair of electrons rather than dots. With this line notation, these two molecules are represented as

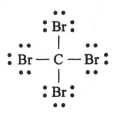

:C ≡ O:

Part B: Electron-Dot Structures Report Sheet

Complete the following table.

Chemical Formula	Total Number of Valence Electrons	Plausible Electron-Dot Structure
HCl		
SiH_4		
PH_3		
O_3		
HOCl		
HCN		
CO_2		
PCl_3		
CH_2Cl_2		
H_2O_2		
H_2CO_2		

CONCEPTUAL INTEGRATED SCIENCE	Activity

The Nature of Chemical Bonds: Polar Bonds and Polar Molecules

Repulsive Dots: VSEPR Theory

Purpose
- To become familiar with the three-dimensional shapes of molecules
- To build molecular models from information given in electron-dot structures
- To draw electron-dot structures from information given in molecular models
- To predict the polarity of a molecule from its molecular shape

Required Equipment and Supplies
molecular model kit
built models of six different molecules

Discussion

It is extremely useful for a chemist to know the three-dimensional shape of the molecules of a given compound. For some compounds (those containing two or more polar bonds), a knowledge of molecular shape is necessary when predicting polarity. One can illustrate the relationship between molecular shape and polarity using water as an example. The O—H bond is polar because of the difference in the electronegativities of the two elements. A water molecule contains two O—H bonds. If a water molecule were linear, as shown in Figure 1, the two polar bonds would be equal in magnitude but opposite in direction. Their effects would cancel, and as a result, the water molecule as a whole would be nonpolar. Water molecules have, however, a bent shape, as shown in Figure 2. In such a structure, the two polar bonds do not cancel, because they do not point in opposite directions. As a result, the water molecule is polar.

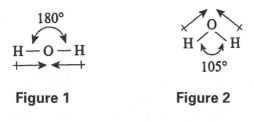

Figure 1	Figure 2

We shall use valence shell electron-pair repulsion theory (VSEPR) as our guide in predicting molecular shapes. The basis of this theory is that pairs of electrons surrounding an atom try to get as far away from one another as possible to lessen electrostatic repulsion. In order to apply this theory, we must know how many pairs of electrons are around each atom in a molecule. It is therefore necessary to begin with the electron-dot structure for the molecule.

Table 1 shows possible molecular shapes derived from VSEPR theory. Note how the electron pairs are placed as far apart as possible, whether the pair is a bonding pair or a nonbonding pair. Molecular *geometry* depends on the total number of substituents about the central atom. A *substituent* is defined as any atom or any nonbonding pair of electrons. Notice that although the total number of electrons around the central atom is always eight (four pairs), the number of substituents can be less than four.

Remember that according to the VSEPR model, pairs of valence electrons strive to get as far apart as possible. With a multiple bond, the pairs cannot separate from one another, because they are all being shared by the same two nuclei. For the purposes of VSEPR, therefore, a multiple bond is treated as a single bond, and the noncentral atom taking part in the bond counts as *one* substituent.

Table 1

General Lewis Structure	Number of Substituents	Number of Nonbonding Pairs	Molecular Geometry	Molecular Shape
B B : A : B B	4	0	109° Tetrahedral	Tetrahedral
B : A : B B	4	1	109° Tetrahedral	Triangular pyramidal
B : A : B	4	2	109° Tetrahedral	Bent
B A :: B B	3	0	120° Triangular planar	Triangular planar
A :: B B	3	1	120° Triangular planar	Bent
B :: A :: B or B : A ::: B	2	0	180° Linear	Linear

Molecular *shape* depends on the placement of *atom substituents only*; nonbonding pairs are *not* considered when you are determining molecular shape. When there are no nonbonding pairs of electrons about a central atom, the molecular shape is identical to the molecular geometry. It is only when nonbonding pairs are present that a molecule's shape differs from its geometry.

Procedure:

Part A: From Electron-Dot Structure to Model
Using the molecular model kit provided by your instructor, build a model of each molecule shown in Table 2 of the report sheet. Then sketch your molecules in the appropriate spaces in Table 2, using the bond notation (—, ········, ◄) used in column 5 of Table 1. Then complete columns 3 and 4 of Table 2.

Part B: From Model to Electron-Dot Structure

Use the three-dimensional models provided by your instructor to complete Table 3 of the report sheet.

Molecular Shapes Report Sheet

Table 2

Electron-Dot Structure	Sketch of Three-Dimensional Model	Name of Shape	Molecular Polar or Nonpolar?
H:N:H (with H below)			
H:C:Cl (with H above and below)			
H:S:H			
:O::C::O:			
:O::S:O:			
H:C:::N:			
H C::O (with H below)			

Table 3

Model Number	Sketch of Three-Dimensional Model	Electron-Dot Structure	Name of Shape	Molecule Polar or Nonpolar?
1				
2				
3				
4				
5				
6				

Summing Up

1. Is BCl_3 a polar substance? Is NCl_3? Explain your answers.

2. Is SCl_2 a bent molecule? Is H_2S? Explain your answers.

Laboratory Manual for *Conceptual Integrated Science,* © 2007 Addison Wesley

Name _____ Date _____

The Nature of Chemical Bonds: Molecular Models

Molecules by Acme

Purpose

In this experiment, you will build models of various molecules. First, you will be given a chemical formula. Then, based upon some rules for how atoms bond, you will piece together a molecular model.

Required Equipment and Supplies

Molecular modeling kits. Students should be able to work in groups of two or three.

Discussion

The three-dimensional shapes of molecules can be envisioned with the use of molecular models. There are certain things you must know, however, before you can assemble an accurate model of a molecule. First, you need to know what types of atoms make up the molecule and also their relative numbers. This information is given by the molecule's chemical formula. The chemical formula for water, H_2O, for example, tells us that every water molecule is made of two hydrogen atoms and one oxygen atom. The second thing you need to know is how the atoms of the molecule fit together. For a water molecule, there are several possibilities. One hydrogen atom might be bonded to both the second hydrogen atom and the oxygen atom (Figure 1a), or all three atoms might be bonded in the shape of a three-member ring (Figure 1b). We find through chemistry, however, that there are specific ways in which different types of atoms bond. Hydrogen, for example, forms only one bond, while oxygen forms two bonds. Knowing this, we can build a more reasonable model of a water molecule where both hydrogen atoms are bonded once to a central oxygen atom (Figure 1c).

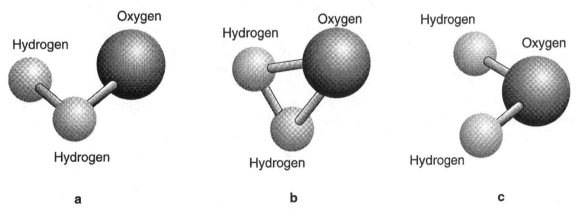

Figure 1. Which version of H_2O is correct?

The shape of a molecule is largely responsible for the physical and chemical properties of the molecule. If water were linear like carbon dioxide, for example, its boiling point would be close to that of carbon dioxide, $-78°C$, which would mean that the Earth's oceans would be gaseous.

The Molecular Model

Molecular modeling kits represent different types of atoms with different colored pieces. Hydrogen atoms, for example, are typically represented by white pieces and oxygen atoms by red pieces. Also, the number of times an atom prefers to bond is indicated by the number of times the piece is able to attach to other pieces. Hydrogen atoms, for example, bond only once, so hydrogen pieces have only one site for attachment. Similarly, oxygen pieces have two sites for attachment. To build a molecule, the pieces representing atoms are connected by sticks (or springs). You will know when you have built a correct structure for a molecule when each atom is bonded the appropriate number of times (Table 1). In some instances, you may find it necessary to form multiple bonds between atoms. Consult the instructions to your modeling kit or your instructor to see how this is done. This is a hands-on, play as you investigate laboratory. Enjoy!

Table 1

Type of Atom	Atomic Symbol	Typical Color of Sphere	Number of Bonds* [holes]
Hydrogen	H	White	1
Carbon	C	Black	4
Nitrogen	N	Blue	3
Oxygen	O	Red	2
Chlorine	Cl	Green	1

*The number of times an atom tends to bond is related to its position in the periodic table. Consider, for example, the relative positions of carbon, nitrogen, oxygen, and chlorine. That these atoms are in adjacent columns and prefer 4, 3, 2, and 1 bonds, respectively, is not a coincidence. The periodic table is more than a table of facts. With further study, you will find that the periodic table is highly organized— and a lot like a road map, for much information about an element can be told merely from its position.

Procedure

Step 1: Get a set of models. You may wish to work with a partner or two. The sets may not contain equal numbers of pieces, so you may occasionally need to borrow from other groups.

Step 2: Determine which colors should be used to represent the following elements: hydrogen, carbon, nitrogen, oxygen, chlorine, and iron. The number of bonds they are able to form should be as listed in Table 1. Enter the actual colors of the pieces you use in Table 2.

Step 3: Build models of each of the following molecules. Avoid forming triangular rings made of three atoms. They are strained and less stable. Check your structure with your teacher—in several cases, there is more than one possibility. Complete Table 3 by drawing an accurate representation of your structure (follow the example of the water molecule). Answer the end-of-activity questions using these models.

1. Hydrogen gas . . . . H_2

2. Oxygen gas . . . . O_2

3. Nitrogen gas . . . . N_2

4. Water . . . . H_2O

5. Hydrogen peroxide . . . . . H_2O_2

6. Ammonia . . . . NH_3

7. Methane . . . . CH_4

8. Dichloromethane. . . . . . . CH_2Cl_2

9. Chloroethanol . . . . C_2H_5ClO

10. Carbon dioxide . . . . CO_2

11. Acetylene. . . . . C_2H_2

12. Ethanol . . . . C_2H_6O

13. Acetic acid. . . . . $C_2H_4O_2$

14. Benzene . . . . C_6H_6

15. Iron (III) Oxide . . . . Fe_2O_3

Laboratory Manual for *Conceptual Integrated Science*, © 2007 Addison Wesley

Table 2

Element:	Hydrogen [H]	Carbon [C]	Nitrogen [N]	Oxygen [O]	Chlorine [Cl]	Iron (III)* [Fe]
Color:						

*For iron, choose a piece that is able to bond in six directions at angles of 90°. The iron will form only three bonds such that three potential bonding sites remain unconnected.

Table 3

Hydrogen	Oxygen	Nitrogen	Water	Hydrogen Peroxide

Ammonia	Methane	Dichloromethane	Chloroethanol C_2H_5ClO	Carbon Dioxide CO_2

Acetylene	Ethanol C H O	Acetic Acid	Benzene C_6H_6	Iron (III) Oxide Fe_2O_3

Summing Up

1. a. Atoms combine to form molecules in specific ratios. In a water molecule, for example, there are two hydrogens for every one oxygen. If this ratio were different—say two hydrogens to two oxygens—would the shape of the molecule also be different?

   _____

   b. Would you still have a water molecule?

   _____

2. a. How are the structures for methane and dichloromethane similar?

   _____

   _____

   b. How are they different?

   _____

   _____

3. How many *different* structures (configurations) are possible for the formula C_2H_5ClO? (Hint: There are more than two.)

   _____

4. Of the 15 models you made, which are linear?

   _____

5. Which molecules have multiple bonds between atoms?

   _____

6. Which of your structures is flat like a pancake?

   _____

 Laboratory Manual for *Conceptual Integrated Science,* © 2007 Addison Wesley

CONCEPTUAL INTEGRATED SCIENCE
Experiment

Chemical Bonds: Separation of a Mixture
Salt and Sand

Purpose
In this experiment, you will develop a laboratory procedure for separating components in a mixture of salt and sand, carry out the separation, and calculate the mass percent of salt and of sand in the mixture.

Required Equipment and Supplies
various salt–sand samples [about 10 grams (g) each]
safety goggles
various pieces of laboratory equipment, depending on the procedure you develop

Discussion
Components of a mixture can be separated by their different physical properties. In this experiment, you are to isolate salt and sand from a mixture. Knowing the mass of the isolated salt and the mass of the mixture, you can calculate the percent composition of the salt that was in the mixture. Likewise, knowing the mass of the isolated sand and the mass of the mixture, you can calculate the percent composition of the sand that was in the mixture. Your instructor will show you how to correctly use and care for laboratory equipment you may use, such as balances and hot plates. The procedure you follow for finding the percent compositions, however, is to be created by you. Before you do anything, therefore, sit down and write out what you think might be a good procedure to follow. For ideas, you might wish to discuss the possibilities with your classmates as well as your instructor. Label this procedure as your "Proposed Procedure."

Before you begin to work in the laboratory, show your written step-by-step proposed procedure to your instructor for final approval. Don't be surprised if you have to modify your procedure as you move along. This is typical. You should also create, therefore, an "Actual Procedure" that shows what was actually done.

Safety!
Significant changes to your proposed procedure must be approved by your instructor. If you have approval to be working with a flame, remove all combustible materials, such as paper towels, from your work station. Eye protection is advised because it is not just what you're doing that poses a hazard. For example, others around you might accidentally shatter glassware, which could send glass pieces toward your face.

Procedure
Step 1: On a separate piece of paper or in a journal, write out a step-by-step procedure that you think will best allow you to find the amount of salt and sand within a mixture. Have your instructor approve of this procedure before moving on. Safety concerns and necessary equipment should be checked. In developing your procedure, keep in mind that you have to end up with the salt and the sand in separate containers so that you can measure the mass of each component.

Step 2: Obtain a sample of salt and sand from your instructor, and write down its reference number by your procedures and within Question 1 of Summing Up.

Step 3: Begin following your proposed procedure, while recording your actual procedure. Write down numerical measurements, such as mass, neatly within one or more data tables. Include units with any number you write down.

Step 4: Use the following equations to calculate the mass percents of salt and sand in your mixture:

$$\text{mass percent of salt} = \frac{\text{mass of salt}}{\text{mass of entire sample}} \times 100$$

$$\text{mass percent of sand} = \frac{\text{mass of sand}}{\text{mass of entire sample}} \times 100$$

Because the mixture contains only two components, the total of the two percentages should be 100%. In reality, various errors may occur such that these two percentages *do not* add up to 100%.

Step 5: Obtain the true values for the mass percents from your instructor.

Summing Up

1. Enter your sample's identification number in this box:

2. What were the mass percents of salt and sand in your sample that you found through your procedure?

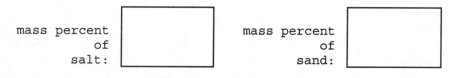

mass percent of salt:

mass percent of sand:

3. What were the true mass percents of salt and sand as reported to you by your instructor after you had completed your procedure?

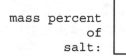

mass percent of salt:

mass percent of sand:

4. What do you think was your most significant source of error in determining the mass percent of salt?

_____

_____

5. What do you think was your most significant source of error in determining the mass percent of sand?

_____

Laboratory Manual for *Conceptual Integrated Science,* © 2007 Addison Wesley

Name _____ Date _____

The Nature of Chemical Bonds: Percent Sugar Determination
Sugar Soft

Purpose
In this experiment, you will determine the sugar content of commercially sold soft drinks using a home-built hydrometer consisting of a 9-inch plastic pipet and nut.

Required Equipment and Supplies
9-inch plastic pipet
1/2-inch nut
sugar solutions (4%, 8%, 12%, 16%)
50-milliliter (mL) graduated cylinder
ruler [centimeter (cm)]
graph paper
various soft drinks (Try fruit juices and diet drinks as well!)

Discussion
A *hydrometer* is a flotation device used to measure the density of a liquid. The greater the density of the liquid, the higher the hydrometer floats. In this exploration, how high the hydrometer floats in four standard sugar solutions will be measured. The greater the sugar content of the solution, the greater its density, hence, the higher the hydrometer floats. A *calibration curve* will be graphed showing the height of the hydrometer on the *y*-axis and the concentration of sugar on the *x*-axis. How high the hydrometer floats in various commercially prepared soft drinks, which are essentially sugar solutions with small amounts of other materials, will then be measured. Using the calibration curve, the sugar content of each soft drink may be estimated.

How high the hydrometer floats out of the water is the distance between its tip and the liquid surface [use units of millimeters (mm)]. Make sure that the hydrometer is not held to the sides of the container—it should float as vertically as possible. The hydrometer must be rinsed and dried before each testing. Also, carbonated beverages must be "decarbonated," because bubbles will collect on the hydrometer and affect its buoyancy. Your instructor will have decarbonated beverages by boiling them and then allowing them to cool.

Procedure

Part A: Construction and Calibration of a Simple Hydrometer
Step 1: Fill the 9-inch pipet about one-half full with water and invert. All the water should run into the bulb.

Step 2: Slip the nut onto the stem end and allow it to rest on the "shoulders" of the bulb as shown to the right.

Step 3: Test your hydrometer by placing it, bulb end down, in the 50-mL graduated cylinder containing 50 mL of water. The pipet should float with about 2.5 cm of the stem sticking out above the water. If it sticks out much more or less than this, either add water to the pipet or remove water from it.

A simple hydrometer

Step 4: When you have adjusted the amount of water in the pipet bulb so that the stem sticks out about 2.5 cm, measure the height of the stem above the surface of the water in the graduated cylinder. Measure to the nearest millimeter, and record the height on Data Table 1.

Data Table 1

Concentration of Sugar Solution (%)	Height of Hydrometer Above Liquid Surface (mm)
0 (plain water)	
4	
8	
12	
16	

Step 5: Remove the hydrometer from the graduated cylinder. *Being careful not to let any water spill out of the hydrometer or any water get in,* rinse the outside surface of the hydrometer well and dry completely.

Step 6: Empty the graduated cylinder, rinse well, and dry completely.

Step 7: Add 50 mL of the 4% sugar solution to the graduated cylinder, place the hydrometer in the solution—bulb end down—and measure and record the height of the stem above the surface of the solution.

Step 8: Repeat Steps 5, 6, and 7 for 8%, 12%, and 16% sugar solutions, and enter all your data into Data Table 1.

Step 9: Plot the information in Data Table 1 into the large graph presented at the end of this exploration. Draw a straight line that best represents all the data points, as the example in Figure 1 shows. This straight line is also known as a calibration curve.

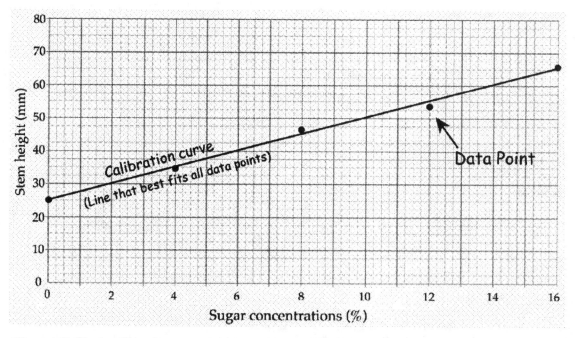

Figure 1. The calibration curve shown here is a line that best represents all of the data points. Note that the calibration curve does not necessarily touch all points.

Laboratory Manual for *Conceptual Integrated Science,* © 2007 Addison Wesley

Part B: Determination of Sugar Content in Beverages

Step 1: Obtain soft drink and/or juice samples from your instructor.

Step 2: Following the procedure given in Part A, measure the stem height for each of your beverages and record this information in Data Table 2.

Step 3: Use the calibration curve you prepared in Part A for your particular hydrometer to determine the sugar concentration in each of your beverages. To do this, find the point on the vertical axis that corresponds to the hydrometer stem height for your beverage. Figure 2 shows an example where the stem height is 52.5 mm. Draw a horizontal line that runs rightward from this point until it intersects the calibration curve. Then, drop a vertical line from the intersection point down to the horizontal axis. The value at the point where this line intersects the horizontal axis—about 10.75% in the example shown here—is the sugar concentration of your beverage.

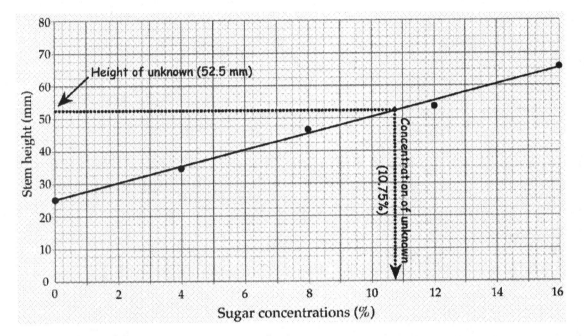

Figure 2. According to this example calibration curve, a stem height of 52.5 mm translates into a sugar concentration of about 10.75%.

Step 4: Record the concentrations of all your beverages in Data Table 2.

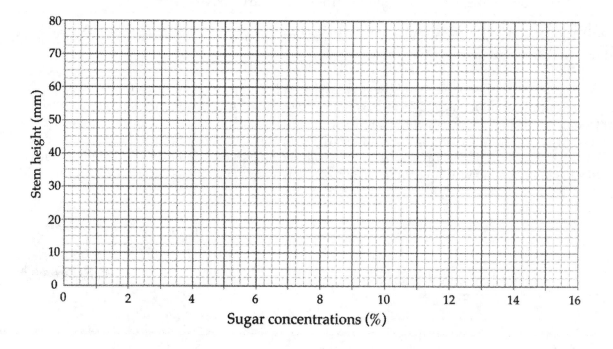

Data Table 2

Brand Name or Type of Beverage	Height of Hydrometer Above Liquid Surface (mm)	Sugar Concentration of Beverage (%)

Laboratory Manual for *Conceptual Integrated Science,* © 2007 Addison Wesley

Name _____   Date _____

| CONCEPTUAL INTEGRATED SCIENCE | Activity |

The Nature of Chemical Bonds: Collection of a Gas
Bubble Round-Up

Purpose
In this activity, you will isolate gaseous carbon dioxide by water displacement.

Required Equipment and Supplies
baking soda
vinegar
2 250-milliliter (mL) Erlenmeyer flasks [or one flask and one 2-liter (L) soda bottle]
1000-mL beaker (or large plastic tub)
2 rubber stoppers (one with glass rod inserted through it)
tubing with a paper clip inserted in one end
ring stand with clamp
safety goggles

Discussion
A common method of collecting a gas produced from a chemical reaction is by the displacement of water. Bubbles of the gas are directed into an inverted flask filled with water. As the gas rises into the container, it displaces water. Once all the water is displaced, the flask may be sealed with a stopper. The physical and chemical properties of the gas can then be investigated. In this activity, baking soda (sodium bicarbonate, $NaHCO_3$), and vinegar (5% acetic acid, CH_3COOH), will be reacted to form gaseous carbon dioxide, CO_2, which will be collected by water displacement. The other products of this reaction—water, H_2O, and sodium acetate, $CH_3COO^-Na^+$—form a liquid phase that remains in the reaction vessel.

$$NaHCO_3 \quad + \quad CH_3CO_2H \quad \Rightarrow \quad CO_2 \quad + \quad H_2O \quad + \quad CH_3CO_2Na$$

| sodium bicarbonate | acetic acid | carbon dioxide | water | sodium acetate |

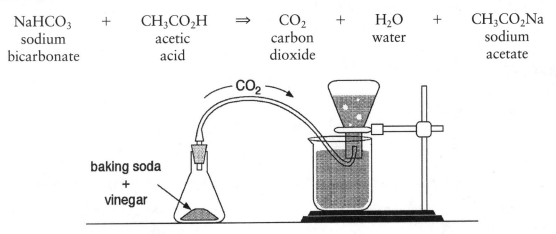

Figure 1. Baking soda (sodium bicarbonate) and vinegar (acetic acid) react to form gaseous carbon dioxide, CO_2.

Procedure
Step 1: Assemble the setup as shown in Figure 1 or one presented to you by your teacher. A 250-mL Erlenmeyer flask is equipped with a rubber stopper through which a glass tube has been inserted. The glass tube is attached to a plastic tube that is forced into a J-shape on the opposite end by the bending of an inserted paper clip. A second 250-mL Erlenmeyer flask is filled with water and inverted into a 1000-mL beaker filled with 700 mL of water. The J-shaped end of the plastic tubing is then fixed below the lip of the inverted flask.

Step 2: Remove the stopper from the upright Erlenmeyer flask and add about a tablespoon of baking soda (sodium bicarbonate) followed by a capful of vinegar (5% acetic acid). Allow the vigorous bubbling to settle down. Add a second capful of vinegar and immediately stopper the upright flask. This will cause bubbles to fill the inverted flask. When bubbling stops, add an additional capful of vinegar to collect additional bubbles (carbon dioxide). Continue in this manner until the inverted beaker has been filled with carbon dioxide. (*Caution: Add vinegar only by the capful. Froth may overflow the flask if you add too much vinegar all at once.*)

Step 3: Stopper the inverted flask tightly to seal the carbon dioxide. Remove the flask from the clamp and turn it right side up with the stopper still in place.

Step 4: Light a wooden match and hold it into the carbon-dioxide-containing flask. The flame should quickly extinguish. You might also light a small candle and slowly pour the carbon dioxide out of the flask onto the candle to extinguish the candle. This works because the carbon dioxide is heavier than air and will thus pour out of the flask.

Going Further

Step 5: Weigh the mass of an empty and dry 2-L bottle with its cap. The mass of the air inside this bottle is about 2.3 grams. Subtract the mass of the air inside the bottle from the mass you measure to get the mass of only the bottle and its cap.

Step 6: Insert the J-hook end of the tube into the upright bottle and pipe lots of carbon dioxide into the bottle. Make sure that the tube is dry and that no water drops get into the bottle. Because the CO_2 is heavier than air, it will settle to the bottom and displace the air, which will escape out the top. Continue filling the bottle with CO_2 until a match is extinguished when placed into the opening.

Step 7: Cap the bottle and measure its mass. Subtract the mass of the bottle and its cap from the mass you measure to get the mass of the carbon dioxide it contains. Calculate the density of the carbon dioxide by dividing its mass by volume. Compare your experimental value to the actual density of carbon dioxide, which is 1.80 grams per liter.

Your calculated density of carbon dioxide:	The actual density of carbon dioxide:
_____	1.80 grams per liter

Summing Up

1. How certain are you that an empty 2-L plastic soda bottle contains a volume of exactly 2 L? How would you find out for sure?

   _____

   _____

2. Why was the flask containing baking soda not stoppered after the first capful of vinegar was added?

   _____

   _____

3. Why was the large beaker not initially filled to the brim?

   _____

   _____

CONCEPTUAL INTEGRATED SCIENCE	Activity

The Nature of Chemical Bonds: Radial Paper Chromatography
Circular Rainbows

Purpose
In this activity, you will separate the different colored components of black ink.

Required Equipment and Supplies
a variety of black felt-tip pens (colored overhead transparency pens also work great)
circular filter paper or chromatography paper (or paper coffee filters)
a variety of solvents:
    water
    methanol (wood alcohol)
    ethanol (grain alcohol)
    isopropyl alcohol (rubbing alcohol)
    white distilled vinegar
    diethyl acetate (fingernail polish remover)
beakers or crucibles upon which to place the paper

Discussion
Black ink is made by combining different colored inks such as blue, red, and yellow. Together, these inks serve to absorb all the frequencies of light. With no light reflected, the ink appears black.

Procedure
It is easy to separate the components of an ink using a technique called *paper chromatography*. Place a concentrated dot of the ink at the center of a porous piece of paper, such as filter paper. Prop the marked paper on a small beaker or crucible, then carefully add a drop of solvent such as water, white vinegar, or rubbing alcohol on top of the dot. Watch the ink spread radially with the solvent. Because the various components have differing affinities for the solvent, they travel with the solvent at different rates. Just before your drop of solvent is completely absorbed, add a second drop, then a third, and so on, until the components have separated to your satisfaction. How the components separate depends on several factors, including your choice of solvent and your technique. Black felt-tip pens tend to work best, but you should experiment with a variety of different types of pens—even food coloring. It is also interesting to watch the leading edge of the moving ink under a microscope. Check for capillary action!

Summing Up
1. The different components of ink not only have differing affinities for the solvent, they also have differing affinities for the paper, which is fairly polar. How might you expect a component that is ionically charged to behave while using a relatively nonpolar solvent, such as acetone—would it readily travel with the solvent or might it stay behind?

    _____

2. Did you ever notice that not all "blue" inks are the same color blue? How might this be explained?

    _____

Name _____ Date _____

The Nature of Chemical Bonds: Purification of Brown Sugar

Pure Sweetness

Purpose
In this activity, you will isolate white sugar from brown sugar.

Required Equipment and Supplies
brown sugar
kitchen cooking pot or 2000-milliliter (mL) beaker
empty glass ketchup bottle or 1000-mL beaker
kitchen knife or microspatula
small widemouthed jar (such as a baby food jar)
safety goggles
gloves for handling heated pot or beaker

Discussion
A supersaturated solution of brown sugar in water may be prepared by dissolving ample amounts of brown sugar in a small amount of boiling water. Upon cooling and after several days, a large number of crystals will have grown. These crystals will still be brown but not quite as brown as the original brown sugar. Recrystallization of these brown crystals will result in crystals that are even less brown. Repeated recrystallization may result in crystals that appear white.

The formation of crystals is almost as much of an art as it is a science. It is difficult, therefore, to guarantee that a particular procedure will always result in crystals of a given size and quality. With this in mind, expect to make some on-the-fly modifications to the following procedure.

This activity may last for several weeks, and you should keep a journal of the equipment and materials that you use and the actual procedure that you follow. Be sure to date every entry. Drawing sketches of your setup is also important.

Procedure
Step 1: In a cooking pot or large beaker, bring about 200 mL of water to a boil. Slowly add brown sugar to the boiling water with continuous stirring. Continue adding brown sugar until it becomes thick and just begins to froth. Do not overcook the brown sugar.

Step 2: Allow the solution to cool before pouring it into an empty glass ketchup bottle or into a 1000-mL beaker. Allow the syrup to stand for several days until a significant number of crystals have formed. The syrupy solution from which the crystals form is known as the *mother liquor*. Your crystals at this point will likely appear as tiny hard clumps.

Step 3: After there has been a significant amount of crystal growth, pour out the mother liquor. You may discard the mother liquor, but first note its close resemblance to molasses. In fact, that's exactly what it is! The brown color results from the many plant by-products it contains. The mother liquor will pour out very slowly. Consider leaving it inverted for several hours. The crystals should stay behind, stuck to the glass walls.

Step 4: Rinse the collected crystals with warm water to remove additional mother liquor. Rinse only briefly, as you want to avoid dissolving these crystals. After this quick rinse, isolate a few of these crystals and examine them closely.

Step 5: Use hot water to help collect all your crystals from your container. Use as little hot water as you can. Transfer your crystals with the small amount of water to a small pot or beaker and apply heat until the crystals dissolve and some frothing is seen. You should end up with a syrupy solution once again.

Step 6: Allow the syrupy solution to cool before transferring it back to the cleaned ketchup bottle or beaker. A glass baby food jar also works well. Wait several days for more crystals to grow. This is called *recrystallization*, because you are crystallizing a material that had already been crystallized. The effect is an increase in the crystal's purity.

Step 7: After crystal growth, the mother liquor of this solution can be poured out. The crystals that remain behind can be collected onto a towel, rinsed briefly with warm water and quickly dried so that they retain crystal shape. If crystals still retain significant brown color, they may be recrystallized again using the general procedures given above.

Summing Up

1. How might you prove that commercial-grade white sugar still contains small amounts of molasses?

_____

_____

2. Which is more *pure:* white sugar or brown sugar?

_____

_____

3. Which is more *natural:* white sugar or brown sugar?

_____

_____

4. Write a statement regarding the quality vs. quantity of the sugar crystals you obtained from this activity.

_____

_____

Name _____ Date _____

Chemical Reactions: Red Cabbage Juice pH Indicator
Sensing pH

Purpose
In this activity, you will isolate and use the pH-sensitive pigment of red cabbage.

Required Equipment and Supplies
red cabbage
cooking pot
strainer
clear plastic cups or 100-milliliter (mL) beakers (5)
safety goggles
water
household solutions: white vinegar, ammonia cleanser, grapefruit juice, club soda, toilet bowl cleaner
household powders: baking soda, washing soda, borax, salt
9-volt battery

Discussion
The pH of a solution can be approximated by way of a pH indicator—any chemical whose color changes with pH. Many pH indicators can be found in plants. Red cabbage is one such plant.

Safety!
In this activity, you will be creating a broth of red cabbage in boiling water. Protect your eyes from the splattering of any liquid or the shattering of any glassware by wearing safety goggles. Protect your skin from scalding by handling the boiling water with care. If you work with toilet bowl cleaner, exercise extra caution, as this material is very acidic.

Procedure
Step 1: Shred about a quarter of a head of red cabbage. Boil the shredded cabbage in about 500 mL of water for about 5 minutes. Strain the cabbage while collecting the broth, which contains the pH indicator. Allow the broth to cool so that it can be safely handled.

Step 2: Fill five or more clear plastic cups or 100-mL glass beakers halfway with the broth. To each cup add a small amount of a household solution (about 10 mL) or powder (about a teaspoon). Examples of solutions you might use include white vinegar, ammonia cleanser, grapefruit juice, club soda, and toilet bowl cleaner. Examples of powders you might use include baking soda, washing soda, borax, and salt.

Step 3: Note the changes in color and estimate the pH of the solution based upon the following:

Color of red cabbage indicator	dark red	pinkish red	light purple	light green	dark green
pH range	1–4	4–6.5	6.5–7.2	7.2–8	8–10

Enter the identity of the material you are testing and the estimated pH of the solution it makes in Table 1.

Step 4: Add some fresh broth to a clear cup and carefully submerge the terminals of a 9-volt battery into this broth. Hold onto the battery. Do not drop it into the broth. Look carefully for any change in color around either of the terminals. If you don't see color changes, you may have to dilute the solution. Look also for the formation of bubbles. Dry the 9-volt battery thoroughly after use and return it to your teacher.

Table 1

Household solution or powder	Estimated pH

Summing Up

1. What color changes did you see upon submerging the terminals of the 9-volt battery into the red cabbage broth?

_____

2. Hydroxide ions, OH^-, are one of the products that form around the terminals of the 9-volt battery dipped into solution. At which terminal (the positive or negative) did these hydroxide ions form?

_____

3. At what terminal on the 9-volt battery (the positive or negative) did you see the formation of bubbles?

_____

Laboratory Manual for *Conceptual Integrated Science,* © 2007 Addison Wesley

| CONCEPTUAL INTEGRATED SCIENCE | Experiment |

Chemical Reactions: Titration

Upset Stomach

Purpose
In this experiment, you will measure and compare the acid-neutralizing strengths of antacids.

Required Equipment and Supplies

Equipment
buret with stand with clamps
three 250-milliliter (mL) Erlenmeyer flasks
safety goggles
gloves
balance
mortar and pestle
weigh dishes
plastic pipets
well-plates

Chemicals
0.50 *M* HCl solution
0.50 *M* NaOH solution
phenolphthalein indicator
various brands of antacid tablets

Safety!
This is a fairly safe lab. Eye protection, however, must be worn at all times when you or your lab partners are working with solutions of hydrochloric acid, HCl, or sodium hydroxide, NaOH. You may also want to wear gloves to protect your skin, which may sting if it comes in contact with the hydrochloric acid or feel slippery if in contact with the sodium hydroxide. If either of these solutions gets onto your skin, tell your teacher and proceed to rinse your skin thoroughly with running water.

Discussion
Overindulging in food or drink can lead to acid indigestion, a discomforting ailment that results from the excess excretion of hydrochloric acid, HCl, by the stomach lining. An immediate remedy is an over-the-counter antacid, which consists of a base that can neutralize stomach acid. In this experiment, you will add an antacid to a simulated upset stomach. Not all the acid will be neutralized, however, and so you will then determine the effectiveness of the antacid by determining the amount of acid that remains.

This is done by completing the neutralization with another base, sodium hydroxide, NaOH. The reaction between hydrochloric acid and sodium hydroxide is

$$\text{HCl} \quad + \quad \text{NaOH} \quad \rightarrow \quad \text{NaCl} \quad + \quad \text{H}_2\text{O}$$

hydrochloric   sodium   salt   water
acid   hydroxide

The concentrations of the HCl and NaOH used in this experiment are the same. This means that the volume of NaOH needed to neutralize the remaining HCl in the "relieved" stomach is equal to the volume of HCl *not* neutralized by the antacid.

Procedure

Part A: Preparing the Upset Stomach
Step 1: Obtain a 250-mL Erlenmeyer flask.

Step 2: Using a buret, deliver about 30.00 mL of a 0.50 M HCl solution to your 250-mL Erlenmeyer flask.

Step 3: With the proper number of decimal places as specified by your instructor, record in your data sheet the volume of 0.50 M HCl that your flask actually contains. For example, depending on the instrument used to deliver the HCl solution, your volume might be 29.93 mL or 29.9 mL or simply 30 mL.

Step 4: Add two drops of phenolphthalein indicator to the flask. No color change should be observed.

Part B: Adding the Antacid
Step 1: Record the brand and active ingredient of your antacid on the data table.

Step 2: Crush and grind the antacid tablet with a mortar and pestle. (Note: Alka-Seltzer need not be crushed.)

Step 3: Carefully transfer all of the resulting powder to a weigh-dish, and then determine and record the mass of the weigh-dish plus powder.

Step 4: Carefully transfer the antacid from the weigh-dish to the upset stomach flask prepared in Part A, swirling for a few minutes while being careful not to spill. This flask now represents your "relieved upset stomach." (The solution should remain colorless.)

Step 5: Determine and record the mass of the empty weigh-dish.

Step 6: Calculate the mass of the antacid tablet.

Part C: Completing the Neutralization
Step 1: Have your instructor show you the proper technique for delivering drops of liquid from a plastic pipet. This involves holding the pipet firmly while gently squeezing the bulb so that you can control drops coming out of the pipet one at a time. Practice your technique by delivering drops of water over a sink.

Step 2: Use a clean pipet to transfer 20 drops of the solution in the "relieved upset stomach" flask to each of five wells in the plastic well-plate.

Step 3: Obtain a small beaker containing about 20 mL of 0.50 M NaOH solution. Using a clean pipet, slowly add this solution *dropwise* to one of the wells containing the "relieved upset stomach" solution. Keep a careful count of the word number of drops added, and continue adding until a light pink color appears. Hold the well-plate to the tabletop and gently swirl to help mix the solutions. The pink color should fade. Continue adding drops and swirling until the light pink color persists for at least 30 seconds. Placing the well-plate on a white sheet of paper will make it easier to see the pink color.

Step 4: Record the number of drops of 0.50 M NaOH added in the data table, and then repeat Step 3 for the remaining wells containing "relieved upset stomach" solution.

Step 5: Repeat the entire procedure, starting from Part A, for other brands of antacid.

Data

Parts A and B: Preparing the Upset Stomach and Adding the Antacid

	Antacid 1	Antacid 2	Antacid 3
1. Antacid used	_____	_____	_____
2. Active ingredient	_____	_____	_____
3. Volume of HCl added to Erlenmeyer flask	_____ mL	_____ mL	_____ mL
4. Number of drops of phenolphthalein added	_____drops	_____drops	_____drops
5. Mass of weigh-dish with crushed antacid	_____ g	_____ g	_____ g
6. Mass of weigh-dish after antacid was transferred	_____ g	_____ g	_____ g
7. Mass of antacid added to "stomach"	_____ g	_____ g	_____ g

Part C: Completing the Neutralization

Antacid 1	Well 1	Well 2	Well 3	Well 4	Well 5		Total
1. Drops of relieved stomach fluid	20	20	20	20	20	=	100
2. Drops NaOH added to complete neutralization						=	
3. Drops of stomach acid neutralized by NaOH						=	
4. Drops of stomach acid						=	

Antacid 2	Well 1	Well 2	Well 3	Well 4	Well 5		Total
1. Drops of relieved stomach fluid	20	20	20	20	20	=	100
2. Drops NaOH added to complete neutralization						=	
3. Drops of stomach acid neutralized by NaOH						=	
4. Drops of stomach acid						=	

Antacid 3	Well 1	Well 2	Well 3	Well 4	Well 5		Total
1. Drops of relieved stomach fluid	20	20	20	20	20	=	100
2. Drops NaOH added to complete neutralization						=	
3. Drops of stomach acid neutralized by NaOH						=	
4. Drops of stomach acid						=	

Summing Up

1. List the antacids you tested in order of neutralizing strength, strongest first:

   strongest _____ > _____ > _____ weakest

2. List the antacids you tested in order of the masses of the tablets, most massive first:

   most massive _____ > _____ > _____ least massive

3. Divide the total number of drops of stomach acid neutralized by each antacid by the mass of the antacid that was added to the stomach:

   Antacid 1: _____ drops/gram

   Antacid 2: _____ drops/gram

   Antacid 3: _____ drops/gram

4. List the antacids in order of neutralizing strength based upon the number of drops of acid neutralized for every gram of antacid:

   strongest _____ > _____ > _____ weakest

5. What would be the effect on the neutralizing strength for an antacid if the following errors were made?

   a. A student uses two tablets of one of the antacids rather than just one tablet:

   _____

   b. A student spills some of the crushed antacid as it is transferred to the "stomach" flask:

   _____

   c. A student "overshoots" the number of drops of sodium hydroxide so that the solution turns a dark pink rather than a light pink:

   _____

Laboratory Manual for *Conceptual Integrated Science,* © 2007 Addison Wesley

CONCEPTUAL INTEGRATED SCIENCE	Activity

Chemical Reactions: Percent Oxygen in Air

Tubular Rust

Purpose
In this activity, you will determine the percent oxygen in air.

Required Equipment and Supplies
steel wool
test tube [15 centimeters (cm)]
pencil
250-milliliter (mL) beaker
centimeter stick and rubber band
white distilled vinegar (5% acetic acid)
ring stand and test tube clamp

Discussion
This activity takes advantage of the rusting of iron by oxygen to determine the percent oxygen in the air. Iron is placed in an air-filled test tube, which is then inverted in water. As the iron reacts with the oxygen, the pressure inside the test tube decreases, and atmospheric pressure pushes water up into the tube. The decrease in volume of air in the test tube is a measure of the depletion of its oxygen content. By measuring the quantity of air in the test tube before and after the rusting of the iron, the percent oxygen in air by volume can be calculated.

Procedure
Step 1: Measure about 1 gram of steel wool that has been pulled apart to increase its surface area. To promote rusting, dip the wool into some white distilled vinegar. Shake off the excess vinegar, and push the steel wool halfway into a test tube 15 cm in length using a pencil. The wool should be packed as loosely as possible to maximize its surface area but packed tight enough so that it doesn't fall out upon inverting the tube.

Step 2: Attach a lightweight centimeter stick to the test tube using a rubber band. Position the stick so that the 0-millimeter (mm) mark is toward the open end. Carefully invert the test tube into the beaker, which should be filled halfway with water. Clamp the test tube to a ring stand and adjust its height so that the lip is no more than a centimeter under the water. Adjust the ruler so that the 0-mm mark is even with the water level *inside* the tube.

Step 3: Water will climb up into the test tube as the rusting proceeds. As this occurs, lower the test tube deeper into the beaker so that the water levels outside and inside the test tube remain even. Begin to read and record the water level inside the test tube at 3-minute intervals until it stops rising.

Step 4: Plot a graph of the water level inside the test tube vs. time.

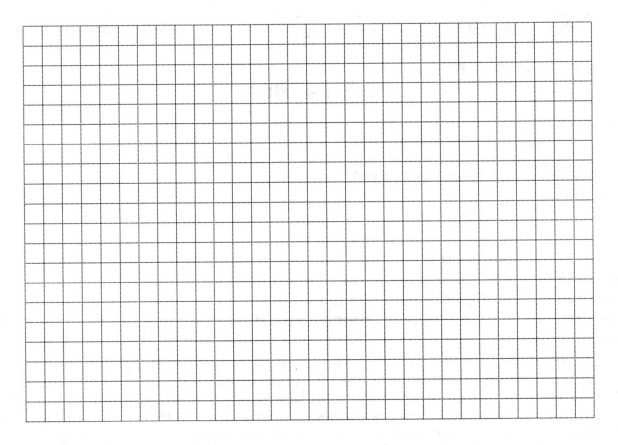

Step 5: Calculate the fraction of oxygen in air as the maximum height of the water inside the tube divided by the total length of the tube. Multiply by 100 to obtain a percentage.

Summing Up

1. What is your experimental percent oxygen in the air?

_____

_____

2. Why is it important that the water levels inside and outside the test tube remain even?

_____

_____

3. Would it take a longer or shorter time for the oxygen to be depleted if the steel wool were packed tightly at the bottom of the test tube? Briefly explain.

_____

_____

Name _____ Date _____

CONCEPTUAL INTEGRATED SCIENCE	Activity

Organic Chemistry: Preparation of Fragment Esters

Smells Great!

Purpose
In this activity, you will prepare a variety of fragrant esters.

Required Equipment and Supplies
watch glasses [5–10 = centimeter (cm) diameters]
concentrated sulfuric acid in dropper bottle
safety goggles
various alcohols and carboxylic acids to be reacted together

Alcohols		Carboxylic Acids
methanol	+	salicylic acid
octanol	+	acetic acid
benzyl alcohol	+	acetic acid
isoamyl alcohol	+	acetic acid
n-propanol	+	acetic acid
isopentenol	+	acetic acid
isobutanol	+	propionic acid

Discussion
Many flavorings and fragrances, both natural and artificial, are a class of organic molecules called *esters*. These molecules contain a carbonyl group bonded to an oxygen atom that is bonded to a carbon atom (see Chapter 14, *Conceptual Integrated Science*). Esters can be prepared in the laboratory by reacting alcohols with carboxylic acids (Figure 1).

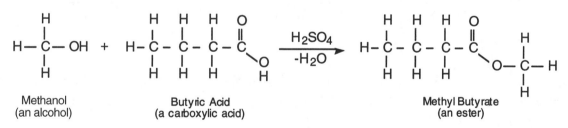

Methanol (an alcohol)   Butyric Acid (a carboxylic acid)   Methyl Butyrate (an ester)

Figure 1. An example of the formation of an ester from an alcohol and a carboxylic acid.

In this activity, you and your class will prepare small but fragrant quantities of different kinds of esters and attempt to identify their smells. To minimize the production of organic wastes, each lab group may be assigned specific esters to prepare.

Safety!
This lab requires the use of concentrated sulfuric acid, H_2SO_4. Only one drop per reaction will be used, but because this acid is concentrated, these drops can be very harmful to skin. Your instructor will be dispensing drops of concentrated sulfuric acid onto your watch plates. Handle your samples with care. They may smell good, but do not taste them, because they may still contain residual amounts of sulfuric acid. Wear safety goggles at all times. When you are finished with this activity, empty your samples into the provided waste container. Rinse your watch glasses with the solvent provided for you by your instructor and then clean them with soap and water.

Procedure

Step 1: Place 1 milliliter (mL) of an alcohol along with 1 mL (or 1 gram, if a solid) of the corresponding carboxylic acid as given in Table 1 onto a watch glass. Smell your samples before adding sulfuric acid. Note: It is improper and unsafe laboratory practice to stick the sample really close to your nose and sniff heavily. Instead, odors can be brought to the nose by waving your hand over the watch glass, which may be held several inches away from your face.

Step 2: To initiate the formation of an ester from the alcohol and the carboxylic acid, bring your samples to the instructor who will add a drop of concentrated sulfuric acid. Gently stir with a microspatula to promote mixing. As the ester forms, a noticeable change in odor will occur. Some samples may take longer than others—allow up to 10 minutes.

Step 3: Complete Table 1 by matching the observed smell with one of the following fragrances. Check with other lab groups for samples that were not assigned to you.

Possible fragrances include apple, banana, "Juicy Fruit," orange, peach, pear, rum, and wintergreen.

Table 1

Alcohol	Carboxylic Acid	Observed Smell
methanol	salicylic acid	
octanol	acetic acid	
benzyl alcohol	acetic acid	
isoamyl alcohol	acetic acid	
n-propanol	acetic acid	
isopentenol	acetic acid	
methanol	butyric acid	
isobutanol	propionic acid	

Summing Up

1. What's the difference between an ester found in a natural product, such as a pineapple, and the same ester produced in the laboratory?

_____

_____

2. Why do foods become more odorous at higher temperatures?

_____

_____

CONCEPTUAL INTEGRATED SCIENCE	Activity

Organic Chemistry: Densities of Organic Polymers

Name That Recyclable

Purpose
In this activity, you will identify a variety of unknown recyclable plastics based on their densities.

Required Equipment and Supplies
recyclable plastics (PET, HDPE, LDPE, PP, PS)
solutions
    95% ethanol and water (1:1 by volume)
    95% ethanol and water (10:7 by volume)
    10% NaCl in water

Discussion
There are many different types of plastics, and many of them are recyclable. Ultimately, plastics need to be sorted according to their composition. For this reason, plastics are coded with a number within the recycling arrow logo. The initials of the type of plastic may also appear. For example, the plastic used to make 2-liter soft-drink bottles is polyethylene terephthalate (PET). This plastic has the following recycling code:

Plastics can also be identified based upon their densities, which is useful if the recycling code is unreadable or absent. In this activity, you are to identify pieces of unknown plastics based upon their densities.

Procedure
Step 1: Use the information in Table 1 to develop a separation scheme by which you will be able to identify all your unknown pieces of plastic.

Step 2: Use the solutions provided to identify each unknown piece of plastic:

Unknown No.:					
Identity:					

Table 1

Recyclable Plastic	Density Compared to...			
	Water	Ethanol and Water (1:1)	Ethanol and Water (10:7)	10% NaCl in Water
1 PET — Polyethylene terephthalate	Greater	Greater	Greater	Greater
2 HDPE — High-density polyethylene	Less	Greater	Greater	Less
4 LDPE — Low-density polyethylene	Less	Less	Greater	Less
5 PP — Polypropylene	Less	Less	Less	Less
6 PS — Polystyrene	Greater	Greater	Greater	Less

Summing Up

1. What other physical properties might be used to identify unknown pieces of plastic?

_____

_____

2. You've just been given a thousand pounds of recyclable polystyrene and polypropylene plastic pieces all mixed together. Suggest how you might quickly separate the different types of plastic from one another.

_____

_____

Laboratory Manual for *Conceptual Integrated Science,* © 2007 Addison Wesley

CONCEPTUAL INTEGRATED SCIENCE | Experiment

The Basic Unit of Life—The Cell: Diffusion
Magnifying Microscopes

Purpose
During this lab, you will learn the parts of the microscope and use one to investigate osmosis in cells.

Required Equipment and Supplies
fresh potato, peeled and sliced
3 bowls
salt
Elodea leaves
microscopes
slides
coverslips
eyedroppers
newspaper
scissors
saltwater
isotonic solution
hypotonic solution
hypertonic solution
variety of prepared slides for viewing

Discussion
All living things are made up of cells. Yet most cells are too small for the human eye to observe without the aid of a microscope. Anton van Leeuwenhoek first described bacteria, yeast, and a variety of other microscopic organisms in the late 1600s. He was able to do this because of improvements he made to the light microscope, which was invented in the 1500s, by a father and son, Zaccharias and Hans Janssen. Since van Leeuwenhoek's time, microscopes have increased in power, but the principle of how they magnify is still the same. In this lab, you will learn the parts of a microscope and how it magnifies.

Next, you will use the microscope to perform an experiment involving isotonic, hypotonic, and hypertonic solutions. *Isotonic* solutions have the same concentration of solutes as a cell. *Hypertonic* solutions have a higher concentration of solutes than does a cell. *Hypotonic* solutions contain a lower concentration of solutes than does of a cell. Water within the cells can freely pass through the cell membrane via *osmosis*. The water moves in an attempt to create equilibrium between the fluid inside and outside the cell.

Procedure
Listen to your teacher as he or she goes through the rules for using a microscope.

Step 1: Examine your microscope, finding each of the parts listed in the picture below.

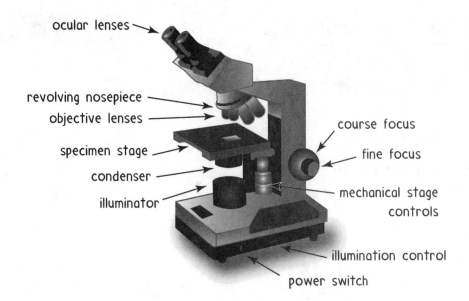

ocular lenses

revolving nosepiece
objective lenses
specimen stage
condenser
illuminator

course focus
fine focus
mechanical stage
controls
illumination control
power switch

Step 2: Calculate the power of magnification for each of the objectives using the formula:

Eyepiece power × objective power = total magnification

Scanning objective: _____   High-Power Objective: _____

Low-Power objective: _____   Oil Immersion Objective: _____

Step 3: Cut a letter "e" from the newspaper, and make a wet mount by placing the "e" on a slide, putting a drop of water on top of it, and placing a coverslip over both. (Note: to avoid air bubbles, carefully place the coverslip down starting with one edge and then gently lower the rest of the coverslip so that air escapes.)

Step 4: Place the slide onto the microscope stage making sure it is placed properly within the clips. Move the stage using the mechanical stage controls so that the "e" is centered under the scanning objective.

Step 5: Bring the letter "e" into focus by slowing raising the stage toward the scanning objective with the course adjustment knob. Once the "e" is mostly in focus, use the fine adjustment knob to bring it into sharp clear focus. Draw what you see.

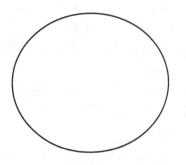

Step 6: Switch to the low-power objective. DO NOT move the course adjustment knob! Using the fine focus adjustment knob, bring the "e" into focus. Draw what you see.

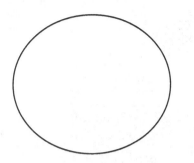

Step 7: Move your stage up and down and to the right and left. Record how the "e" appeared as it moved.

_____

_____

Step 8: Switch to the high-power objective. DO NOT move the coarse adjustment knob! Using the fine focus adjustment knob, bring the "e" into focus. Draw what you see.

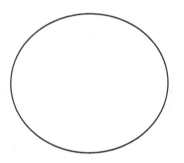

Step 9: Switch your microscope back to the scanning objective, and view one of the prepared slides. Practice bringing the slide into focus on the scanning objective and then switching to low power and high power. Remember—DO NOT move the coarse adjustment knob after you get the slide in focus with the scanning objective. Record what you see. Be sure to label your drawings.

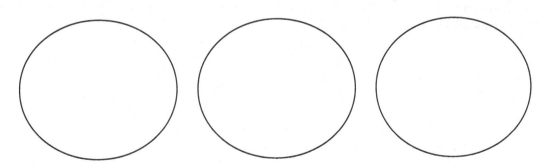

Step 10: Obtain three _Elodea_ leaves, and make a wet mount of each using the following. Slide 1: use the isotonic solution; Slide 2: use the hypotonic solution; and Slide 3: use the hypertonic solution. Let slides sit for at least 10 minutes.

Step 11: Make a prediction about what will happen to each leaf. Be as specific as possible.

Isotonic Solution	Hypertonic Solution	Hypotonic Solution

Step 12: While you are waiting to observe your *Elodea* slide take two slices of potato. Observe the slices. Place one slice in a bowl with saltwater, and the other in a bowl of fresh tap water. Predict what will happen to each. Be specific.

Saltwater Slice	Pure Water Slice

Step 13: Observe each of the *Elodea* slides using the low and high objectives. Record what you observe. Be sure to label your drawings.

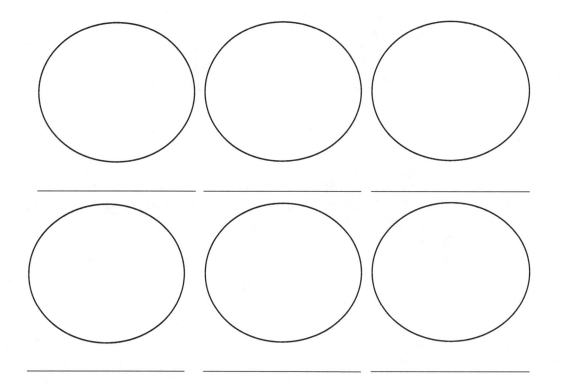

Laboratory Manual for *Conceptual Integrated Science,* © 2007 Addison Wesley

Step 14: Observe your potato slices, and record your observations:

Potato Slice in Saltwater	Potato Slice in Pure Water

Summing Up

1. How did the "e" appear to move when you moved the stage in Step 7?

_____

_____

_____

2. What happened to the *Elodea* leaf in the hypertonic solution? Explain why. _____

_____

_____

3. How was the *Elodea* leaf in the hypotonic solution different? Why?

_____

_____

_____

4. Based on the potato slice experiment, explain what happens when you eat potato chips?

_____

_____

5. What other experiment could we have done to illustrate the difference between isotonic, hypertonic and hypotonic, solutions? _____

_____

_____

CONCEPTUAL INTEGRATED SCIENCE	Activity

Genetics: Inheritance

Real-Life Inheritance

Purpose
Through this lab, you will learn about real-life inheritance and get the opportunity to investigate one genetic disorder in depth.

Required Equipment and Supplies
PTC paper
Internet/library access

Discussion
Is real-life inheritance as simple as Mendelian genetics? In some cases, yes—for example, whether a child is a boy or girl is determined by two sex chromosomes (XX for a girl, XY for a boy). But unknown to Gregor Mendel, the human genome is composed of chromosomes, and on those chromosomes are genes. Genes located on the same chromosome are said to be *linked,* meaning that during meiosis, they will be kept together and inherited as a unit. Other traits are the result of multiple genes, which is known as *pleiotropy.* Still other traits are determined by two alleles that come from a pool of more than two choices.

Procedure
Activity 1: Mendelian Characteristics in Humans

Step 1: Go through each item in the table below, and determine your phenotype and genotype. (Note: If you have the dominant phenotype, you can only put down T* for the genotype, because you know only that you have one dominant allele—the second could be dominant or recessive.)

Trait	Phenotype	Genotype
Sex		
Thumb Hyperextensibility: dominant (T*) Straight: recessive (tt)		
Earlobe Free: dominant (F*) Attached: recessive (ff)		
Hairy knuckles Hair on middle phalanx: dominant (K*) No hair: recessive (kk)		
Hairline Widows peak: Dominant (H*) Straight hairline: recessive (hh)		
PTC paper Tastes bitter: Dominant (P*) No taste: recessive (pp)		

Activity 2: Sex-Linked Traits

Some traits are sex-linked, which means that the genes controlling the trait are located on the sex chromosome. For example, color blindness is a sex-linked trait. In order to have color blindness, a male must have inherited one X chromosome containing the recessive gene. In contrast, a color-blind female must inherit two X chromosomes with the color-blind gene. If the female inherits one X chromosome containing the color-blind gene, she is considered a *carrier*. X_B is normal (no color blindness), X_b contains the color-blind gene.

Normal mom ($X_B X_B$) has a child with a color-blind dad ($X_b Y$)

Possible gametes from mom: X_B, X_B

Possible gametes from dad: X_b, Y

	X_B	X_B
X_b	$X_B X_b$	$X_B X_b$
Y	X_B Y	X_B Y

Phenotype ratio: 1:1 (50% female color-blind carriers, 50% normal males)

Step 1: Hemophilia is also a sex-linked disease (X_h). Complete a punett square for a normal mom and hemophiliac dad.

List the possible gametes for mom: _____

List the possible gametes for dad: _____

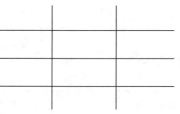

Step 2: Cross a hemophiliac mom with a normal dad.

Step 3: Cross a hemophiliac mom and dad.

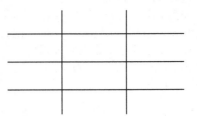

Activity 3: Genetic Diseases
Step 1: Choose a genetic disease to research.

Step 2: Research the disease, finding out what type of genetic disorder it is and five other specific facts about the disease. Record your findings in the "Summing Up" section below.

Summing Up
Activity 1: Mendelian Characteristics in Humans
1. Based on Activity 1, could you predict your parents' phenotype for hairline? Explain.

_____

_____

_____

2. Would it be possible for your parents both to have hairless knuckles and you to have hairy knuckles? Explain. _____

_____

_____

Activity 2: Sex-Linked Traits
1. If a mom has hemophilia, what is the probability that her children will have an affected X chromosome? Explain. _____

_____

_____

2. If a dad has hemophilia, what is the probability that his children will have an affected X chromosome? Explain. _____

_____

_____

3. If a mother who is a carrier of hemophilia has a child with a normal dad, what is the probability that their child will also have hemophilia? Explain.

_____

_____

4. After doing this activity, would you say that men or women are more susceptible to sex-linked diseases? Why is this so? Explain. _____

_____

_____

Activity 3: Genetic Diseases

1. What genetic disorder did you research? _____

2. What type of genetic disorder is it (i.e., autosomal recessive, dominant, trisomy, sex-linked)?

_____

3. List five facts you learned about the disease.

   1. _____

       _____

   2. _____

       _____

   3. _____

       _____

   4. _____

       _____

   5. _____

       _____

4. List your sources.

   1. _____

       _____

   2. _____

       _____

   3. _____

       _____

CONCEPTUAL INTEGRATED SCIENCE	Activity

The Evolution of Life: Charles Darwin and The Origin of Species
Understanding Darwin

Purpose
Through the activities in this lab, you will have the opportunity to explore Darwin's theory of evolution.

Required Equipment and Supplies
plastic spoons
paper clips

Discussion
While many scientists have contributed to the theory of evolution, Charles Darwin is the scientist most often associated with the theory. He is credited with coming up with the following key ideas about evolution:

1. *Descent with modification*: Populations change over time and adapt to their environments. Sometimes, this descent with modification, or *evolution,* results in the creation of new species over time (a process known as *speciation*).

2. *Common descent*: Organisms originated from a common ancestor. This idea is supported by similarities in structure (anatomy), embryonic development, and genetics. (Note that Darwin did not have the extensive knowledge of genetics that we have today.)

3. *Natural selection*: Because there are not enough resources for all organisms to survive and to reproduce as much as they can, living organisms are involved in an intense "struggle for existence." As a result, organisms with advantageous traits leave more offspring than organisms with other traits, causing populations to change over time.

Procedure
Activity 1: What Is Natural Selection?
For this first activity, you will play a series of games with your class to help you understand how natural selection works.

Step 1: When you think of the phrase "natural selection" what skills or characteristics come to mind first? _____

Step 2: The scenario: You are on an island, and the only way to obtain food is to climb a cliff, which requires arm strength. Arm wrestle with your classmates to see which of you can obtain food from the cliff. Once you lose a round, you are considered dead. Continue the contest until there are only four competitors left. Record their names on the board.

Step 3: A second scenario: You are on an island where you must reach down into a poisonous plant to obtain its nutritious, tasty fruit. The catch? You cannot touch the sides of the plant, because if you do, you will immediately drop dead from poison. To test how steady your hand is, you will be given a plastic spoon and paper clips. You must pile ten paper clips onto the spoon and then walk across the room as fast as you can without dropping any. If you drop any paper clips, you are eliminated. Record the names of the four fastest people on the board.

Step 4: Discuss what you think would happen if you had to climb the cliff and also reach into the plant to obtain food (in other words, if both skills were needed).

_____

_____

Step 5: In small groups, discuss a wild animal you are familiar with and list all of the characteristics it has to help it survive.

_____   _____   _____

_____   _____   _____

Step 6: Answer the "Summing Up" questions at the end of the lab.

Activity 2: How Does Natural Selection Work?
Step 1: Record everyone's shoe size on the board. Create a bar graph from the data to use as a reference.

Step 2: Interpret the data:

What is the range of shoe sizes? _____

Which shoe size is most common? _____

Step 3: Answer the questions in the "Summing Up" section below.

Activity 3: How Does Speciation Occur?
Step 1: The scenario: your class is a population surviving on an island. One day, a terrible storm goes through the island, separating it in half. Half the students are stranded on each side of the island with no possibility of contacting the other half. After you are assigned to Half A (the mountainous side of the island) or Half B (the open, flat part of the island) by your instructor, think about what characteristics you will need to survive on your half of the island and list them below:

_____

_____

_____

_____

Step 2: Discuss what adaptations may develop in your population over the next three millennia, and list them below:

_____

_____

_____

_____

Step 3: After many generations go by, the two halves of the island are reconnected, making contact possible. Gather as a class and share your adaptations with each other. Then answer the questions in the "Summing Up" section below.

Summing Up

Activity 1: What Is Natural Selection?
1. Were the same four people surviving at the end of both games? If so, why do you think this happened?

_____

_____

_____

2. What did you anticipate would happen if you needed to be able to climb the cliff and reach into the plant for food? _____

_____

_____

3. Based on the games you played, define natural selection in your own words.

_____

_____

_____

Activity 2: How Does Natural Selection Work?
1. Did your class show much variation in shoe size?

_____

_____

_____

2. What are the advantages to having variation within a population? _____

_____

_____

3. Think about a real-world scenario in which variation within a population plays an important role, and write your thoughts below. _____

_____

_____

Activity 3: How Does Speciation Occur?

1. Based on your discussion in class, do you think that the two halves of the island would eventually form different species? Explain.

_____

_____

_____

2. List two other ways a new species could evolve.

a. _____

b. _____

Name _____ Date _____

The Evolution of Life: Evidence of Evolution

Investigating Evolution

Purpose

In this lab, you will review the theory of evolution by investigating scientific evidence for evolution and how scientists obtain this evidence.

Required Equipment and Supplies

Internet/library access

Discussion

The theory of evolution is a *scientific theory,* based on a large amount of data. But how was this data obtained? There are no records from the Paleozoic era. We cannot repeat the development of ancient organisms in a laboratory, nor can we know the exact meteorological conditions of the past. So how do scientists know what they do about evolution?

Procedure

Step 1: Based on what you know, list all of the techniques and evidence you can think of that scientists use to support the theory of evolution.

_____

_____

_____

Step 2: Share your ideas with the class, and compile a list of techniques and evidence on the board. Your instructor will then assign your group one of these techniques or lines of evidence.

Step 3: Research your assigned topic using the Internet and library as resources. Specifically, your group should research the details and history of your technique or line of evidence and provide three specific examples of how it provides evidence for the theory of evolution.

Step 4: Write up your research in the "Summing Up" section below, and be prepared to present it in the class as a news report. Cite your research references in the area below.

Summing Up

Research findings: _____

_____

_____

_____

_____

_____

_____

_____

_____

_____

_____

_____

_____

References:

1. _____

2. _____

3. _____

4. _____

5. _____

CONCEPTUAL INTEGRATED SCIENCE	Experiment

Biological Diversity: Classifying Living Things

What Is It—Bacterium? Protist? Fungi?

Purpose
This lab will help you understand how scientists classify organisms and the challenges they face. You will also learn how to identify unclassified organisms.

Required Equipment and Supplies
microscopes
fresh pond water
clean slides
coverslips
eyedropper

Discussion
The group bacteria is composed of single-celled prokaryotes—that is, organisms that lack a nucleus. Bacteria are believed to have been the first living organisms on earth, with a history stretching back over 3.5 billion years. Bacteria outnumber all eukaryotes combined—in fact, one handful of fertile soil contains more bacteria than the number of people who have ever lived on earth! In today's society, bacteria sometimes get a bad rap as disease-causing agents, but the fact is that less than 1% of bacteria are pathogens. Bacteria come in various shapes, including bacilli (rods), cocci (spheres), and spirochetes (spirals). They can be found as individuals or in pairs, chains, or small colonies. All bacteria reproduce via asexual reproduction.

The protists include a wide variety of eukaryotes—organisms whose cells contain a nucleus. Most protists are unicellular, but some form colonies or multicellular structures. Protists include heterotrophs, such as amoebas and ciliates, and autotrophs, such as dinoflagellates, diatoms, and green algae.

Fungi are eukaryotes. Most are multicellular, although there are a few notable exceptions such as yeast. In many ecosystems, fungi play critical roles as decomposers, recycling organic molecules from dead organisms. Fungi are heterotrophs and obtain their food via absorption. They reproduce sexually as well as asexually through the release of spores.

Procedure
Step 1: Describe the pond water.

What color is it? _____

Is it clear or cloudy? _____

What does it smell like? _____

Can you see any living organisms? _____

How many organisms do you expect to find? _____

Step 2: Make a slide of pond water by placing several drops of pond water on a clean slide. Cover the drop with a coverslip.

Step 3: View the slide under the microscope. Start at scanning power. Bring the slide into focus, and record what you see by drawing a picture and writing a brief description.

Step 4: View the slide under the microscope. Move to low power. Bring the slide into focus, and record what you see by drawing a picture and writing a brief description.
(Hint: Many of the organisms you are looking for are nearly translucent, so keep the light levels low, and move your slide back and forth as you try to bring it into focus.)

Step 5: View the slide under the microscope. Move to high power. Bring the slide into focus, and record what you see by drawing a picture and writing a brief description.

Step 6: Using the chart below, try to identify as many organisms you observed as you can and record your findings in the table.

Category	Organism Name
Bacteria	
Protist	
Fungi	
Plant	
Animal	

Summing Up

1. How many different organisms did you find in your sample of pond water? Did you find the same organisms as the groups around you?

_____

_____

2. What types of organisms did you see—bacteria? fungi? protists? plants? animals? Which category had the most organisms? _____

_____

_____

3. Describe how you were able to identify the organisms in the pond water. Did you look at their shape? Their size? How they moved?

_____

_____

CONCEPTUAL INTEGRATED SCIENCE	Activity

Human Biology I—Control and Development: The Skeleton and Muscles

Muscles and Bones

Purpose
In this lab, you will learn the major bones of the skeletal system and the major muscles of the muscular system.

Required Equipment and Supplies
model skeleton
muscle magazines
scissors
glue
paper
markers
old textbooks or Web images with labels to use as reference material

Discussion
We cannot see our skeletal system, yet without it, we would be like giant blobs with no real form. We would not be able to move, because our muscles would have no place to attach to or push and pull against. Many internal organs, including our brain, spinal cord, heart, lungs, and liver, would be left unprotected and vulnerable to damage. Our circulatory and immune systems would have no bone marrow to make blood cells. So the bottom line is: even though we cannot see our skeleton, we need it to survive.

Likewise, our muscles are essential to life in countless ways. We all have three types of muscle: cardiac, smooth, and skeletal. Without our cardiac muscle, we would have no heart to pump blood. Without our smooth muscle, the food we eat could not be digested to provide an ongoing supply of nutrients for us to live on. Without skeletal muscle, we could not do any of those voluntary activities that make life enjoyable, like go for a bike ride, eat a sundae, or move our eyes to read a book.

Procedure
Activity 1: Skeletal System
Step 1: Discuss with your lab group strategies for learning the major bones in the body in the next 20 minutes. Record the strategy you are going to use. _____

_____

Step 2: Get to work learning your bones using the available resources provided by your instructor. Give yourselves just 20 minutes.

Step 3: Test how much you learned by labeling the parts of the skeleton below with the correct names.

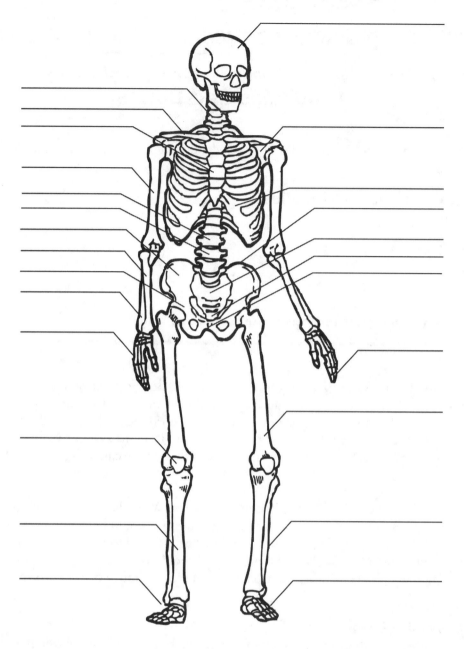

Activity 2: Muscle System

Step 1: Cut out pictures of various muscles in the body from the muscle magazines provided by your instructor. Be sure you have pictures of the front and back of the head, torso, arms, and legs.

Step 2: Glue the pictures to the paper provided, leaving enough room around each picture for labels.

Step 3: Discuss a strategy with your group for learning the names of muscles in the picture.

Step 4: Get to work on learning the names using the resources provided by your instructor. Give yourselves 20 minutes.

Laboratory Manual for *Conceptual Integrated Science,* © 2007 Addison Wesley

Step 5: Test your knowledge of muscles by labeling each of the following on your picture:

Masseter

Sternocleidomastoid

Spinalis

Rectus abdominus

External oblique

Trapezius

Serratus anterior

Deltoid

Pectoralis major

Latissimus dorsi

Biceps brachii

Triceps brachii

Brachioradialis

Gluteus maximus

Hamstrings

Sartorius

Quadriceps

Gracilis

Tibialis anterior

Gastrocnemius

Soleus

Achilles' tendon

Obicularis oculi

Obicularis oris

Frontalis

Summing Up

1. To move your humerus, which muscles would be involved?

_____

_____

2. List the bones and muscles that are involved in walking. _____

_____

_____

3. List the muscles involved in each of the following activities:

Wrinkling your forehead _____

Winking your eye _____

Puckering your lips _____

Name _____ Date _____

Human Biology II—Care and Maintenance: The Circulatory System

Keep Pumping

Purpose
In this lab, you will learn to identify the parts of the heart and explore how heart rate changes with your need for oxygen.

Required Equipment and Supplies
stopwatch
pencil
place to run

Discussion
The circulatory system is vital for survival. It can never rest. On average, the heart beats over 100,000 times a day! The circulatory system is responsible for supplying every cell in the body with oxygen and nutrients and taking away carbon dioxide and other wastes. It also circulates the white blood cells that make up the immune system. Bottom line—without a properly functioning cardiovascular system, life comes to an end.

You can think of the heart and attached vessels as a plumbing system in a building. The strength of the pump determines not only the amount of water pumped, but also the water pressure at the destination. Varying the sizes of the pipes also influences water pressure and how hard the pump has to work to get water through.

In order for blood to flow through the circulatory system, the heart must pump hard enough to generate sufficient pressure. Each time the ventricles of the heart contract, they create a pulse of pressure. We can figure out the amount of pressure by taking a blood pressure measurement. We can also figure out heart rate by measuring pulse rate. Pulse rate can be measured at any large artery, but typically, the pulse is taken either at the wrist, right below the thumb, or at the side of the neck, 3 inches below the earlobe. Pulse rate and blood pressure are influenced by many factors, including oxygen need, health, body position, and age. The last two parts of this lab will examine some of these factors.

Procedure
Activity 1: Wrist Pulse
Step 1: Start by locating the pulse on the inside of the forearm near the wrist. Using any finger EXCEPT the thumb, gently feel for the pulse on the thumb side of the wrist.

Step 2: Both you and your lab partner measure your own wrist pulse while sitting quietly. Count the number of pulses felt for 30 seconds and then multiply by two. Record the results in the table below.

Steps 3–5: Repeat the procedure three additional times, and record each result in the table below.

	Own Wrist Pulse: Trial 1	Own Wrist Pulse: Trial 2	Own Wrist Pulse: Trial 3	Own Wrist Pulse: Trial 4
Lab Partner 1				
Lab Partner 2				

Step 6: Calculate the average wrist pulse for each of you. (To calculate an average, add the four pulse rates and divide by four.)

_____ (Average for Lab Partner 1)   _____ (Average for Lab Partner 2)

Steps 7–10: Repeat the procedure, but this time, take your lab partner's wrist pulse while he or she sits quietly. Record the results in the table below.

	Partner's Wrist Pulse: Trial 1	Partner's Wrist Pulse: Trial 2	Partner's Wrist Pulse: Trial 3	Partner's Wrist Pulse: Trial 4
Lab Partner 1				
Lab Partner 2				

Step 11: Calculate the average wrist pulse for each of you. (To calculate an average, add the four pulse rates and divide by four.)

_____ (Average for Lab Partner 1)   _____ (Average for Lab Partner 2)

Activity 2: Neck Pulse (Carotid Artery Pulse)

Step 1: Start by placing your second and third fingers approximately 3 inches below your earlobe. This should put you on or near the carotid artery. Locate the pulse in this area.

Step 2: Both you and your lab partner take your own carotid pulses while sitting quietly. Count the number of pulses felt for 30 seconds and then multiply by two. Record this result in the table below.

Steps 3–5: Repeat the procedure three additional times, and record each result in the table below.

	Own Carotid Pulse: Trial 1	Own Carotid Pulse: Trial 2	Own Carotid Pulse: Trial 3	Own Carotid Pulse: Trial 4
Lab Partner 1				
Lab Partner 2				

Step 6: Calculate your average carotid pulse. (To calculate an average, add the four pulse rates and divide by four.)

_____ (Average for Lab Partner 1)   _____ (Average for Lab Partner 2)

Steps 7–10: Repeat the procedure, but this time, take your lab partner's carotid pulse while he or she sits quietly. Record the results in the table below.

	Partner's Carotid Pulse: Trial 1	Partner's Carotid Pulse: Trial 2	Partner's Carotid Pulse: Trial 3	Partner's Carotid Pulse: Trial 4
Lab Partner 1				
Lab Partner 2				

Laboratory Manual for *Conceptual Integrated Science,* © 2007 Addison Wesley

Step 11: Calculate the average carotid pulse for each of you. (To calculate an average, add the four pulse rates and divide by four.)

_____ (Average for Lab Partner 1)   _____ (Average for Lab Partner 2)

Activity 3: Changes of Pulse Rate

Step 1: Start by vigorously running in place for 1 minute. Have your lab partner take your pulse for 1 minute IMMEDIATELY after you stop running. (Use either wrist or carotid pulse.) Record the pulse rate in the table below.

Steps 2–4: Run for another minute, and have your lab partner count and record your pulse rate as instructed above. Repeat the procedure two more times, having your lab partner take your pulse after each 1-minute run. Record the results in the table below after each minute.

Step 5: Cool down. Continue to have your partner take your pulse every minute until your pulse returns to within five beats of your resting pulse. Record the results in the table below.

Step 6: Switch roles, and repeat the above procedure.

Running Pulse	Pulse After 1-minute Run	Pulse After 2-minute Run	Pulse After 3-minute Run	Pulse After 4-minute Run
Lab Partner 1				
Lab Partner 2				

Cool-Down Pulse Rate	After 1 min	After 2 min	After 3 min	After 4 min	After 5 min	After 6 min	After 7 min	After 8 min	After 9 min	After 10 min
Lab Partner 1										
Lab Partner 2										

Activity 4: Regulating Pulse Rate

Step 1: Discuss with your partner what you could do other than exercise to raise and lower your heart rate.

Step 2: Try out at least two ideas, and record your findings. _____

_____

_____

_____

Summing Up

Activity 1: Wrist Pulse

1. What could account for the similarities or differences between the wrist pulses of you and your lab partner? _____

_____

_____

2. Explain why the pulse rate may have varied between trials. _____

_____

_____

3. Was there a difference between the pulse rates when taken by self vs. partner? Explain.

_____

_____

Activity 2: Neck Pulse (Carotid Artery Pulse)

1. Which was easier to feel—the carotid pulse or the wrist pulse? Explain your answer.

_____

_____

_____

Activity 3: Changes of Pulse Rate

1. Why did your pulse rate change over the exercise period? _____

_____

_____

2. Did you and your partner show the same change in heart rate? Explain why. _____

_____

_____

3. How long did it take your heart rate to return to normal? _____

4. How did that compare to your partner? What could account for the differences or similarities?

_____

_____

_____

Laboratory Manual for *Conceptual Integrated Science,* © 2007 Addison Wesley

Activity 4: Regulating Pulse Rate

1. Why might being able to control your heart rate be a useful skill to have? _____

_____

_____

CONCEPTUAL INTEGRATED SCIENCE	Activity

Human Biology II—Care and Maintenance: Respiration

Breathe In, Breathe Out

Purpose
In this lab, you will explore the respiratory system.

Required Equipment and Supplies
balloons
straws
tape measures
paper bags
cups of water

Discussion
When was the last time something took your breath away? In your lifetime, you will take approximately 470 million breaths. Breathing allows you to take in oxygen, and oxygen allows your cells to make adenosine triphosphate (ATP), without which they would quickly die. Not only does your respiratory system take in oxygen, but it adjusts how much oxygen it takes in according to your body's needs.

Procedure
Activity 1: Control of the Respiratory System
Step 1: Breathe quietly for 3 minutes, then hold your breath following a normal inhalation. Continue holding your breath for as long as possible. Record the time in the table below.

Step 2: Breathe quietly for 3 minutes. Place one end of a straw in your mouth and the other in a glass of water, then hold your breath following a normal inhalation. Just before you have to breathe, start to take a sip of water. Continue holding your breath for as long as possible. Record the time in the table below.

Step 3: Rest until your breath is back to normal. Hyperventilate (take short, quick breaths) for 10 seconds, and then hold your breath following a normal inhalation. Continue holding your breath for as long as possible. Record the time in the table below.

Step 4: Rest until your breath is back to normal, and then place a paper bag over your mouth and nose. Hyperventilate for 10 seconds into the bag, and then hold your breath following a normal inspiration. Continue holding your breath for as long as possible. Record the time in the table below.

Step 5: Run in place for 2 minutes, and then hold your breath. Continue holding your breath for as long as possible. Record the time in the table below.

	Normal Breathing	Straw	Hyperventilation	Paper Bag	Running
Time					

Activity 2: Lung Capacity

Step 1: Standing with your back straight, stretch a balloon, and then, breathing normally, blow up the balloon as much as you can with one breath.

Step 2: Measure the balloon with the tape measure from the end you blew in all the way around back to the end. Record the measurement below.

_____ mm

Step 3: Get into position as though you are performing a stomach crunch. Stretch a new balloon, and breathing normally, blow up the balloon as much as you can with one breath.

Step 4: Measure the balloon with the tape measure from the end you blow in all the way around back to the end. Record the measurement below.

_____ mm

Step 5: Record your answer on the board with the rest of the class. Include with your measurement whether you have a cold, are an athlete, or suffer from asthma or another respiratory illness.

Activity 3: Respiratory Rate—The Effect of Activity

Step 1: Count the number of breaths you take in 1 minute while resting. Record the number in the table below.

Step 2: Count the number of breaths you take in 1 minute *immediately after* jumping in place for 2 minutes. Record the number in the table below.

Step 3: Count the number of breaths you take in 1 minute *5 minutes after* jumping in place. Record the number in the table below.

Step 4: Wait until your breath has returned to normal. While holding your breath, jump in place for 1 minute, then count the number of breaths in 1 minute *immediately after* you jump. Record the number in the table below.

Step 5: Count the number of breaths in 1 minute *5 minutes after* jumping in place while holding your breath. Record the number in the table below.

	Rest	Immediately After Jumping	5 minutes After Jumping	1 minute After Holding Breath Jumping	5 minutes After Holding Breath Jumping
Number of breaths in 1 minute					

Laboratory Manual for *Conceptual Integrated Science,* © 2007 Addison Wesley

Summing Up
Activity 1: Control of the Respiratory System
1. What gases are inhaled and exhaled during normal breathing? _____

2. Under which circumstance could you hold your breath for the shortest amount of time? The longest? Explain the physiological reasons for this. _____

_____

_____

3. Was there a difference in the length of time you could hold your breath after hyperventilating vs. hyperventilating into the paper bag? Explain why. _____

_____

_____

Activity 2: Lung Capacity
1. Was there a difference in lung capacity when you were standing straight as compared to being in a sit-up position? Explain what could account for the difference or similarity. _____

_____

_____

2. As you look over your class's data, did you notice any big differences in the results? Give at least three possible reasons for the differences you saw. _____

_____

_____

Activity 3: Respiratory Rate—The Effect of Activity
1. How did your breathing rate change after jumping in place for 1 minute? _____

_____

_____

2. Explain why your breathing rate changed. _____

_____

_____

3. Did you feel different while you were jumping while holding your breath? Explain. _____

_____

_____

4. Was your breathing rate different 1 minute after jumping while holding your breath compared to 1 minute after jumping while breathing? Explain. _____

_____

_____

Laboratory Manual for *Conceptual Integrated Science,* © 2007 Addison Wesley

CONCEPTUAL INTEGRATED SCIENCE	Activity

Rocks and Minerals: The Formation of Minerals
Minerals and Their Uses

Crystal Growth

Purpose
In this activity, you will observe the growth of crystals from a melt and from a solution.

Required Equipment and Supplies
thymol
sodium chloride
sodium nitrate
potassium aluminum sulfate (alum)
copper acetate
glass slide or clear glass plate
forceps
petri dish
hot plate
microscope

Discussion
A **mineral** is a naturally formed inorganic solid composed of an ordered array of atoms. The atoms in different minerals are arranged in their own characteristic ways. This systematic arrangement of atoms is known as mineral's crystalline structure, which exists throughout the entire mineral specimen. If crystallization occurs under ideal conditions, it will be expressed in perfect crystal faces.

Growth of Crystals from a Melt
Most minerals originate from magma. Crystalline minerals form when molten rock cools. The temperature at which minerals crystallize is very high (well above the boiling point of water), making direct observation difficult. In this experiment, we will use thymol, an organic chemical that crystallizes near room temperature. Although thymol is different from magma, the crystallization principles exhibited are similar.

Safety!
Thymol is not poisonous but may cause skin and eye irritation. Be safe: use forceps to handle thymol.

Procedure
Part A: Slow Cooling Without "Seed" Crystals
Set the hot plate to low heat. Place a petri dish containing a small amount of crystalline thymol on the hot plate until all the crystals are melted. Allow the melt to heat for 1 to 2 minutes and then set it aside to cool slowly. Do not disturb it during cooling. This melt will be examined at the end of the lab session.

Part B: Slow Cooling with "Seed" Crystals

Repeat procedure (A) but transfer the petri dish to the stage of a microscope as soon as the thymol is melted. Add several (four or five) "seed" crystals to the melt and observe it under the microscope. As the melt cools, you will see the crystals begin to grow.

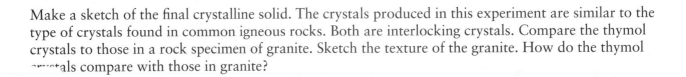

Describe the manner of crystal growth. Think about the rate of growth, the direction of growth, the crystal faces, and the effect of limited space on crystal shape. What role do the "seed" crystals play in initiating crystal growth?

_____

_____

Make a sketch of the final crystalline solid. The crystals produced in this experiment are similar to the type of crystals found in common igneous rocks. Both are interlocking crystals. Compare the thymol crystals to those in a rock specimen of granite. Sketch the texture of the granite. How do the thymol crystals compare with those in granite?

_____

_____

Part C: Rapid Cooling

Repeat procedure (A) but transfer the petri dish to the top of an ice cube and let it cool for 30 seconds. The melt will cool very rapidly. Quickly move the petri dish to the microscope and observe the nature of the crystal growth. Time permitting, repeat this procedure until you are sure of your observations and can state your general conclusions.

_____

_____

_____

Part D: Examination of Crystals from Procedure Part A

Examine the thymol crystals from the first part of this exercise. Observe the crystal size. What effect does the rate of cooling have on the crystal size?

_____

_____

Growth of Crystals from Solution

Many minerals are precipitated from aqueous solutions by evaporation. Crystal growth of such minerals can be observed in the laboratory by evaporating prepared concentrated solutions.

Procedure

Obtain concentrated solutions of sodium chloride, sodium nitrate, potassium aluminum sulfate (alum), and copper acetate from your instructor.

Place a drop of each solution on a microscope slide. Label each slide accordingly. As the water evaporates from the slide, crystals of each compound will appear. With time and continued evaporation, the crystals will grow larger.

Summing Up

1. How does crystallization from an aqueous solution compare to crystallization from a melt?

_____

_____

_____

2. Do the crystals precipitated from a solution have unique crystal forms?

_____

_____

_____

3. According to your observations in this activity, do you think that crystal forms provide a good reference for mineral identification?

_____

_____

_____

4. According to your observations in this activity, what is the relationship between and crystal size?

_____

_____

_____

CONCEPTUAL INTEGRATED SCIENCE	**Activity**

Rocks and Minerals: Classifying Minerals
Minerals and Their Uses: Identification of Minerals

What's That Mineral?

Purpose

In this activity, you will observe the physical properties of various samples and identify them by a systematic procedure.

Required Equipment and Supplies

mineral collection
hardness set—piece of glass, steel knife, copper penny,
streak plate (nonglazed porcelain plate)
dilute hydrochloric acid (HCl)

Discussion

A mineral is a naturally formed inorganic solid with a characteristic chemical composition and crystalline structure. The different combinations of its elements and arrangement of atoms determine the physical properties of a mineral: its shape, the way it reflects light, its color, hardness, and its mass.

I. **The physical properties dependent on a mineral's chemical composition include luster, streak, color, specific gravity, and reaction with acid.**

    The **luster** of a mineral is the appearance of its surface when it reflects light. Luster is independent of color; minerals of the same color may have different lusters, and minerals of the same luster may have different colors. Mineral luster is classified as either metallic or nonmetallic.

Test for Luster

Metallic minerals are usually	Nonmetallic minerals are usually
a. gold, silver or black	a. not gold or silver
b. shiny, polished	b. shiny to dull
c. opaque	c. transparent, translucent, or opaque
d. always have a streak	d. rarely have a streak

The **streak** of a mineral is the color of its powdered form. We can see a mineral's streak by rubbing it across a nonglazed porcelain plate. Minerals with a metallic luster generally leave a dark streak that may be different from the color of the mineral. For example, the mineral hematite is normally reddish-brown to black but always streaks red. Magnetite is normally iron-black but streaks black. Limonite is normally yellowish-brown to dark brown but always streaks yellowish-brown. Minerals with a nonmetallic luster either leave a light streak or no streak at all.

Test for Streak

1. Scrape the mineral across a nonglazed porcelain plate.

2. Blow away excess powder.

3. The color of the powder is the streak.

Although **color** is an obvious feature of a mineral, it is not a reliable means of identification. When used with luster and streak, color can sometimes aid in the identification of metallic minerals. Nonmetallic minerals may occur in a variety of colors or be colorless. Therefore, color is not used for the identification of nonmetallic minerals. In this exercise, we will only differentiate between light-colored and dark-colored minerals.

Specific gravity (s.g.) is the ratio of the mass (or weight) of a substance to the mass (or weight) of an equal volume of water. Metallic minerals tend to have a higher specific gravity than nonmetallic minerals. For example, the metallic mineral gold (Au) has a specific gravity of 19.3, whereas quartz (SiO_2), a nonmetallic mineral, has a specific gravity of 2.65.

The **reaction to acid** is an important chemical property often used to identify carbonate minerals. Carbonate minerals effervesce (fizz) in dilute hydrochloric acid (HCl). Some carbonate minerals react more readily with HCl than others. For instance, calcite ($CaCO_3$) strongly effervesces when exposed to HCl, but dolomite [$CaMg(CO_3)_2$] doesn't react unless it is scratched and powdered.

II. **The physical properties dependent on a mineral's crystalline structure include hardness, cleavage and fracture, crystal form, striations, and magnetism.**

The resistance of a mineral to being scratched (or its ability to scratch) is a measure of the mineral's **hardness.** The varying degrees of hardness are represented on Mohs scale of hardness. For this activity, we are concerned with the hardness of some common objects.

Cleavage and fracture are useful guides for identifying minerals. **Cleavage** is the tendency of a mineral to break along planes of weakness. Planes of weakness depend on crystal structure and symmetry. Some minerals have distinct cleavage. Mica, for example, has perfect cleavage in one direction and breaks apart to form thin, flat sheets. Calcite has perfect cleavage in three directions and breaks to produce rhombohedral faces that intersect at 75 degrees. A break other than along cleavage planes is a **fracture.**

Mohs Scale of Hardness

Mineral	Scale Number	Common Objects
Diamond	10	
Corundum	9	
Topaz	8	
Quartz	7	Steel file
Feldspar	6	Window glass
Apatite	5	Pocket knife
Fluorite	4	
Calcite	3	Copper penny
Gypsum	2	Fingernail
Talc	1	

Test for Hardness

1. Place a glass plate on a hard flat surface.

2. Scrape the mineral across the glass plate.

3. If the glass is scratched, the mineral is harder than glass.

4. If the glass is not scratched, the mineral is softer than glass.

5. If the mineral is softer than glass, test to see if it is harder than a copper penny. Scrape the mineral across the penny.

6. If the mineral is softer than the penny, test to see if it is harder than your fingernail. Try to scratch the mineral with your fingernail.

Every mineral has its own characteristic **crystal form.** Some minerals have such a unique crystal form that identification is relatively easy. The mineral pyrite, for example, commonly forms as intergrown cubes, while quartz commonly forms as six-sided prisms that terminate in a point. Most minerals, however, do not exhibit their characteristic crystal form. Perfect crystals are rare in nature because minerals typically grow in cramped, confined spaces.

CLEAVAGE PATTERNS

Number of Cleavage Directions	Shape	Number of Flat Surfaces	Sample Illustration
1	Flat sheets	2	
2 at 90°	Rectangular cross-section	4	
2 not at 90°	Parallelogram cross-section	4	
3 at 90°	Cube	6	
3 not at 90°	Rhombohedron	6	
Fracture	Irregular shape	0	

Some minerals have grooves on their cleavage planes. These grooves, called **striations,** can be used to differentiate between feldspar minerals. Plagioclase feldspars have straight parallel striations on one cleavage plane. Orthoclase feldspar have lines that resemble striations but are actually color variations within the mineral. These grooves are not straight, and they are not parallel to each other. Instead, these "striations" make a criss-cross pattern.

Some minerals exhibit magnetism. To test for magnetism, simply expose the mineral to a small magnet or a compass.

So we see that physical properties that depend on a mineral's crystalline structure include hardness, crystal form, cleavage and fracture, striations, and magnetism.

MINERALS WITH METALLIC LUSTER

Streak Color	Properties	Mineral
Black to gray	silver gray 3 directions of cleavage at 90° cubic crystals hardness = 2.5 specific gravity = 7.6	Galena
Black to gray	black to dark gray magnetic hardness = 6 specific gravity = 5	Magnetite
Black to gray	gray to black marks paper feels greasy hardness = 1–2 specific gravity = 2.2 golden yellow may tarnish purple hardness = 4 specific gravity = 4.2	Graphite Chalcopyrite
Black to greenish black	brass yellow may tarnish green cubic crystals striations hardness = 6 specific gravity = 5.2	Pyrite (fool's gold)
Reddish brown to black	silver to gray may tarnish reddish brown hardness = 5–6 specific gravity = 5	Hematite
Yellowish brown to reddish brown	brown to silver gray to golden brown may tarnish yellowish brown hardness = 5.5 specific gravity = 4	Limonite

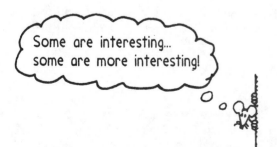

Some are interesting...
some are more interesting!

Laboratory Manual for *Conceptual Integrated Science,* © 2007 Addison Wesley

MINERALS WITH NONMETALLIC LUSTER—DARK COLORED

Hardness	Cleavage	Properties	Mineral
Harder than glass		black to blue gray 2 directions of cleavage, not quite at 90° striations on one cleavage plane hardness = 6 specific gravity = 2.7	Plagioclase
		dark green to black 2 directions of cleavage at nearly 90° no striations hardness = 6 specific gravity = 3.5	Pyroxene
		dark green to black 2 directions of cleavage intersecting at 60° and 120° no striations hexagonal crystals hardness = 5.5 specific gravity = 3.6	Amphibole (hornblende)
Softer than glass	Present	brown to dark green to black 1 direction to cleavage producing thin sheets translucent hardness = 3 specific gravity = 2.9	Biotite
		green to dark green 1 direction of cleavage producing thin curved sheets greasy luster hardness = 2.5 specific gravity = 2.9	Chlorite

Earth science
is down to earth!

MINERALS WITH NONMETALLIC LUSTER—DARK COLORED			
Hardness	Cleavage	Properties	Mineral
Harder than glass	Absent	olive green to black no streak glassy luster conchoidal fracture hardness = 7 specific gravity = 2.6	Olivine
		light to dark gray no streak glassy, transparent, translucent conchoidal fracture hexagonal crystals hardness = 7 specific gravity = 2.6	Quartz
		deep red to brown no streak translucent conchoidal fracture isometric crystals hardness = 7 specific gravity = 4	Garnet
Variable		red reddish brown streak opaque, earthy uneven fracture hardness = 1.5–5.5 specific gravity = 5.2	Hematite

When I dream of geology I have rocks in my head!

MINERALS WITH NONMETALLIC LUSTER—LIGHT COLORED			
Hardness	Cleavage	Properties	Mineral
Harder than glass		pale orange-pink, white	Orthoclase feldspar
		green, brown 2 cleavage directions at nearly 90° no striations color lines on cleavage planes hardness = 6 specific gravity = 2.6	
		white to blue-gray 2 directions of cleavage not quite at 90° striations on one cleavage plane hardness = 6 specific gravity = 2.7	Plagioclase feldspar
Softer than glass	Present	colorless to white 3 directions of cleavage at 90° soluble in water hardness = 2.5 specific gravity = 2.4	Halite
		colorless to white 1 direction of cleavage hardness = 2 (easily scratched with fingernail) specific gravity = 2.3	Gypsum
		colorless to white to yellow 3 directions of cleavage not at 90° (rhomb shaped) translucent to transparent strong reaction to acid hardness = 3 specific gravity = 2.7	Calcite

Hmmm...gold maybe?

MINERALS WITH NONMETALLIC LUSTER—LIGHT COLORED			
Hardness	**Cleavage**	**Properties**	**Mineral**
Softer than glass	Present	white, gray, pink 3 directions of cleavage not at 90° (rhomb shaped) opaque reacts to acid when powdered hardness = 3.5 specific gravity = 2.9	Dolomite
		yellow, blue, green, violet 4 directions of cleavage transparent to translucent cubic crystals hardness = 4 specific gravity = 3.2	Fluorite
		colorless to pale green 1 direction of cleavage producing thin elastic sheets hardness = 2.5 specific gravity = 2.7	Muscovite
		white to greenish 1 direction of cleavage pearly luster hardness = 1 specific gravity = 2.8	Tale
Harder than glass	Absent	white, gray, pink, violet glassy luster conchoidal fracture hardness = 7 specific gravity = 2.6	Quartz
		olive green to yellow green glassy luster conchoidal fracture hardness = 7 specific gravity = 3.5	Olivine

Minerals...sigh!

 Laboratory Manual for *Conceptual Integrated Science,* © 2007 Addison Wesley

Procedure

Examine the various unknown minerals and note their characteristics on the following worksheet. Identify the different minerals by comparing your list to the mineral identification tables.

Separate metallic and nonmetallic minerals.

A. If mineral is metallic determine:
1. streak
2. color
3. hardness
4. any other distinguishing properties
5. the name of the mineral

B. If the mineral is nonmetallic determine:
1. color (separate light minerals from dark minerals)
2. hardness
3. cleavage
4. any other distinguishing properties
5. the name of the mineral

Mineral Identification Sheet						
Luster	Streak	Color	Hardness	Cleavage	Other Characteristics	Mineral Name

Summing Up

1. What distinguishing characteristic is used in identifying the following minerals?

   a. halite _____

   b. pyrite _____

   c. quartz _____

   d. biotite _____

   e. fluorite _____

   f. garnet _____

2. Indicate whether the following physical properties result from a mineral's crystalline structure or a mineral's chemical composition.

   a. crystal form _____

   b. color _____

   c. cleavage _____

   d. specific gravity _____

3. If a mineral does not exhibit a streak, is it metallic? Explain.

   _____

   _____

4. Which property is more reliable in mineral identification, color or streak? Why?

   _____

   _____

5. What physical properties distinguish biotite from muscovite?

   _____

   _____

6. What physical properties distinguish plagioclase feldspars from orthoclase feldspars?

   _____

   _____

Laboratory Manual for *Conceptual Integrated Science,* © 2007 Addison Wesley

CONCEPTUAL INTEGRATED SCIENCE	Activity

Rocks and Minerals

Rock Hunt

Purpose
In this activity, you will practice finding the geology all around you.

Required Equipment and Supplies
desire

Discussion and Procedure

Activity 1
Go on a rock hunt. Now that you are more familiar with rocks and minerals, you can start your own collection. Gather rocks from the beach, and old river channel, a stream bed, a road cut, or even your backyard. Collect at least six different looking rocks and classify them into the three major rock groups. What tell-tale features help you classify the rocks? Can you identify your rocks by these features?

Activity 2
Rocks are found not only on the beach and in the mountains, but almost everywhere. In fact, if you live in a city you are probably surrounded by more rock materials than you realize. Take a field trip down any city street and you will notice that most buildings are constructed from stone material. Many stone buildings are polished, providing an easy view of a rock's mineral composition, texture, and hence its method of formation. Is the rock composed of visible crystals? Are the crystals interlocking? Are the crystals flattened or deformed? What is the grain size? Are there fossils? Support your classification with your observations.

Activity 3
Observe the buildings in your locality. Look for the older buildings, most of which were built from local material. The different rock types used in the construction of some to these buildings can tell you much about local history. Is there a marble or granite quarry in your area? Try to trace the building material to its origin.

CONCEPTUAL INTEGRATED SCIENCE	Activity

Rocks and Minerals: Identification of Rocks

What's That Rock?

Purpose

In this activity, you will identify rocks from the three different types: igneous, metamorphic, and sedimentary.

Required Equipment and Supplies

collection of assorted igneous, sedimentary, and metamorphic rocks
dilute hydrochloric acid (HCl)

Discussion

The three major classes of rocks—igneous, sedimentary, and metamorphic—have their own distinct physical characteristics. By learning to identify the representative characteristics of each rock type, we may gain a better understanding of the history recorded in the earth's crust.

Procedure

You will be given three sets of rocks—igneous, sedimentary, and metamorphic. Your task is to identify the rocks in each set. The first set to identify consists of igneous rocks. Refer to Table 1, *Classification of Igneous Rocks,* to aid in your identification. The second set to identify consists of sedimentary rocks. For this part of the exercise refer to Table 2, *Classification of Sedimentary Rocks.* To make identification and classification easier, the table has been divided into two parts—Part A for clastic sedimentary rocks, and Part B for nonclastic sedimentary rocks. The final set to identify consists of metamorphic rocks. Refer to Table 3, *Classification of Metamorphic Rocks,* to help distinguish the different metamorphic rocks.

Part A: Identification of Igneous Rocks

Molten magma welling up from within the earth produces igneous rock of two types—*intrusive* and *extrusive*. Intrusive rocks are formed from magma that solidified below the earth's surface, and extrusive rocks are formed from magma erupted at the earth's surface.

Step 1: The first step in identifying igneous rocks is to observe their texture. Texture is related to the cooling rate of magma in a rock's formation. Magma that solidifies below the earth's surface cools slowly, forming large, visible interlocking crystals that can be identified with the unaided eye. This coarse-grained texture is described as **phaneritic.** If the texture is exceptionally coarse grained (visible minerals larger than your thumb), the rock is described as having a **pegmatitic** texture. Some intrusive rocks contain two distinctly different crystal sizes in which some minerals are conspicuously larger than the other minerals. This texture, with big crystals in a finer groundmass, is called **porphyritic.**

In contrast, magma that reaches the surface tends to cool rapidly forming very fine-grained rocks. This fine-grained texture is described as **aphanitic.** When magma is cooled so quickly that there is not time for the atoms to form crystals, the texture is described as **glassy.** Many aphanitic rocks contain cavities left by gases escaping from the rapidly cooling magma. These gas bubble cavities are called vesicles, and the rocks that contain them are said to have a **vesicular** texture. If the gaseous magma cools very quickly, the texture that develops is described as **frothy** (foam-like glass). In a volcanic eruption, the forceful escape of gases causes rock fragments to be torn from the sides of the volcanic vent. These rock fragments combine with volcanic ash and cinders to produce a **pyroclastic,** or fragmental texture.

Table 1

Classification of Igneous Rocks				
	Composition			
	Light-Colored	Intermediate Color	Dark-Colored	Very Dark Color
	10–20% quartz K-feldspar > plagioclase ≈ 10% ferromagnesian minerals	No quartz plagioclase > K-feldspar 25–40% ferromagnesian minerals	No quartz plagioclase ≈ 50% 50% ferromagnesian minerals	100% ferromagnesian minerals
Texture Pegmatitic (very coarse grained)	Pegmatite			
Porphyritic (mixed crystal sizes)	Porphyry			
Phaneritic (coarse grained)	Granite Granodiorite	Diorite	Gabbro	
Aphanitic (fine grained)	Rhyolite Peridotite	Andesite	Basalt	Periodite Dunite
Glassy		Obsidian		
Porous (glassy, frothy)	Pumice		Scoria	
Pyroclastic (fragmental)	Volcanic tuff (fragments < 4 mm) Volcanic breccia (fragments > 4 mm)			

Step 2: Observe the color and hence the chemical composition of an igneous rock sample. Igneous rocks are either rich in silicon or poor in silicon. Silicon-rich (quartz-rich) rocks tend to be light in color, whereas silicon-poor rocks tend to be dark in color. Igneous rocks can therefore be divided into a light-colored group—quartz, feldspars, and muscovite—and a dark-colored group—the ferromagnesian minerals—biotite, pyroxene, hornblende, and olivine. [*Ferromagnesian* refers to minerals that contain iron (ferro) and magnesium (magnesian)].

Part B: Identification of Sedimentary Rocks

Rock material that has been weathered, subjected to erosion, and eventually consolidated into new rock is sedimentary rock. Sedimentary rocks are classified into two types—clastic and nonclastic. **Clastic** sedimentary rocks are formed from the compaction and cementation of fragmented rocks. **Nonclastic** sedimentary rocks are formed from the precipitation of minerals in a solution. This chemical process can occur directly, as a result of inorganic processes, or indirectly, as a result of biochemical reaction.

Step 1: Observe the texture of the rock. If the rock is composed of visible particles, the rock is probably clastic. Clastic sedimentary rocks are classified according to particle size. Large-sized particles range from boulders to cobbles to pebbles. If the large-sized particles are rounded, the rock is a **conglomerate.** If they are angular and uneven, the rock is a **breccia.** Medium-sized particles produce the various types of sandstone. **Quartz sandstone** is composed wholly of well-rounded and well-sorted quartz grains. **Arkose** is composed of quartz with about 25% feldspar grains. **Graywacke** is composed of both quartz and feldspar grains with fragments of broken rock in a clay-type matrix. Silts and clays comprise the fine-sized particles. Silt-sized particles produce **siltstone,** and clay-sized particles produce **shale.**

Step 2: If the rock does not fit into the clastic classification, the rock is probably nonclastic. Nonclastic sedimentary rocks are classified by their chemical composition. Inorganic processes produce the **evaporites**—minerals formed by the evaporation of saline water. Evaporites include **rock salt** and **gypsum.** Rock salt can be identified by its salty taste. Biochemical reactions produce **carbonates** by way of calcareous-secreting organism, and **chert,** by silica-secreting organisms. **Limestone, chalk,** and **dolostone** are carbonate rocks. Carbonate minerals react (fizz) to HCl. Chert can be identified by its waxy luster and conchoidal fracture. **Coal** (a biogenic sedimentary rock) is formed from the accumulation and compaction of vegetation matter.

Table 2

Classification of Clastic Sedimentary Rocks			
Particle size		Rock Name	Characteristics
Coarse (>2 mm)		Conglomerate	Rounded grain fragments
		Breccia	Angular grain fragments
Medium (1/16–2 mm)	Sandstone	Quartz	Quartz grains (often well rounded, well sorted)
		Arkose	Quartz and feldspar grains (often reddish color)
		Graywacke	Quartz grains, small rock fragments, and clay minerals (often grayish color)
Fine (1/256–1/16 mm)		Siltstone	Silt-sized particles, surface is slightly gritty
Fine (<1/256 mm)		Shale	Clay-sized particles, surface has smooth feel, no grit

Classification of Nonclastic Sedimentary Rocks			
Composition		Rock Name	Characteristics
Halite (NaCl)	Evaporites	Rock Salt	Salty taste
Gypsum ($CaSO_4 \cdot 2H_2O$)		Gypsum	Inorganic precipitate
Organic matter		Coal	Compacted, carbonized plant remains
Calcium carbonate ($CaCO_3$)		Limestone	Fine grained, strong reaction to HCl
		Chalk	Microfossils; fine grained
		Fossiliferous limestone	Macrofossils and fossil fragments
Dolomite [$CaMg(CO_3)_2$]		Dolostone	Reaction to HCl when in powdered form
Quartz (SiO_2)		Chert	Hard, dense, waxy luster, may show conchoidal fracture

Part C: Identification of Metamorphic Rocks

Rock material that has been changed in form by high temperature or pressure is metamorphic rock. Metamorphic rocks are most easily classified and identified by their texture, and when a particular mineral is very obvious, by their mineralogy. Metamorphic rocks can be divided into two groups: **foliated** and **nonfoliated.**

Foliated metamorphic rocks have a directional texture and a layered appearance. The most common foliated metamorphic rocks are slate, phyllite, schist, and gneiss.

Slate is very fine grained and is composed of minute mica flakes. This most noteworthy characteristic of slate is its excellent rock cleavage—the property by which a rock breaks into plate-like fragments along flat planes. **Phyllite** is composed of very fine crystals of either muscovite or chlorite that are larger than those in slate, but not large enough to be clearly identified. Phyllite is distinguished from slate by its glossy sheen. **Schists** have a very distinctive texture with a parallel arrangement of the sheet-structured minerals (mica chlorite, and/or biotite). The minerals in a schist are often large enough to be easily identified with the naked eye. Because of this, schists are often named according to the major minerals in the rock (biotite schist, staurolite-garnet schist, etc.). **Gneiss** contains mostly granular, rather than platy minerals. The most common minerals found in gneiss are quartz and feldspar. The foliation in this case is due to the segregation of light and dark minerals rather than alignment of platy minerals. Gneiss has a composition similar to granite and often is derived from granite.

Nonfoliated rocks are mono-mineralic and thus lack any directional texture. Their texture can be described as coarsely crystalline. Common nonfoliated metamorphic rocks are **marble** and **quartzite.**

Table 3

Classification of Metamorphic Rocks		
Foliated Metamorphic Rock		
Crystal Size	Rock Name	Characteristics
Very fine, crystals visible	Slate	Excellent rock cleavage
Fine grains, crystals not visible	Phyllite	Well-developed foliation; glossy sheen
Coarse texture Crystals visible with unaided eye Micaceous minerals Often contains large crystals	Schist — Muscovite schist / Chlorite schist / Biotite schist / Garnet schist / Staurolite schist / Kyanite schist / Sillimanite schist	Mineral content reflects increasing metamorphism from top to bottom
Coarse	Gneiss	Banding of light and dark minerals
Nonfoliated Metamorphic Rock		
Precursor Rock	Rock Name	Characteristics
Quartz sandstone	Quartzite	Interlocking quartz grains
Limestone	Marble	Interlocking calcite grains

Laboratory Manual for *Conceptual Integrated Science,* © 2007 Addison Wesley

Summing Up

1. Did you find evidence that igneous rocks can exhibit both fine and coarse textures?

_____

_____

_____

_____

2. What are the influencing factors for large crystal formation? What types of rock exhibit enlarged crystals?

_____

_____

_____

3. What distinguishing characteristics are exhibited in sedimentary rocks?

_____

_____

_____

4. How can we distinguish igneous rocks from metamorphic rocks?

_____

_____

_____

Going Further

For this exercise you will first determine if the rock is igneous, sedimentary, or metamorphic. Then you will use the classification tables for the different rock types to identify the rocks. Examine your rock specimen closely. Look at its texture. Can you see individual mineral grains? If so, the texture is coarse to medium coarse. If the mineral grains are too small to be identified, the texture is fine. How are the grains arranged? Are the grains interlocking? Interlocking grains are formed by crystallization, so the rock is probably either igneous or metamorphic. Are the interlocking grains aligned? Do they show foliation? Foliation indicates metamorphism. Are the grains separated by irregular spaces filled with cementing material? Are fossils present? Does the rock react with acid? If so, the rock is sedimentary. If the rock is crystalline and shows no reaction to acid, check the hardness of the rock. Metamorphic and igneous rocks are harder than sedimentary rocks.

Name _____ Date _____

Earth's Surface—Land and Water: Topographic Maps

Top This

Purpose

In this activity, you will interpret maps, particularly topographic maps, which show landforms and approximate elevations above sea level. Using points of known elevation, you will learn to draw contour lines. Using a topographic map, you will learn to construct a topographic profile.

Required Equipment and Supplies

topographic quadrangle map (provided by your instructor)
ruler, pencil and eraser

Discussion

Maps provide a representation of the earth's surface and are a very useful tool. A map is a scaled-down, idealized representation of the real world. Everything on a map must be proportionally smaller than what it really is. Roads, waterways, mountains, ground area, and distance are all proportionally reduced in scale (Figure 1). A map's **scale** is defined as the relationship between distance on a map and distance on the ground. There are three ways to describe scale on a map. A *graphical scale* is a drawing, a line marked off with distance units. A *verbal scale* uses words to describe distance units— "one inch to one mile." And a *representative fraction* (rf) gives the proportion as a fraction or ratio, such as 1:24,000. This means that one unit of measure on the map—1 inch or 1 centimeter—is equal to 24,000 units of the same measure on the ground. If the scale is 1 : 10,000, then 1 inch on the map is equal to 10,000 inches on the ground. The first number refers to the map distance and is always 1. The second number refers to ground distance and will change depending on scale.

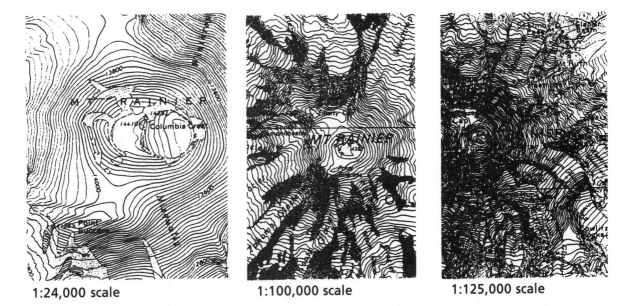

1:24,000 scale   1:100,000 scale   1:125,000 scale

Figure 1. U.S. Geological Survey Maps of the same area at different scales. The scales of a map tells us how much area is being represented and the level of detail in the area. When comparing maps, we see that the smaller the scale the larger the area represented, and the larger the scale the smaller the area represented.

Large-Scale Map Example: 1:10,0000	Small-Scale Map Example 1:1,000,000
Shows a small area with a lot of detail. A large scale is good for urban, street, or hiking maps that require detail.	Shows a large area with very little detail. A small scale is good for world or regional maps that cover a large area.

Scale Conversion

It is difficult to envision 10,000 inches let alone 1,000,000 inches, on the ground. In order to make these units more meaningful, we can convert them into miles or kilometers.

Question: One inch on a 1:25,000 scale represents what distance on the ground?

Answer: Using conversion

1 foot = 12 inches
5280 feet = 1 mile
1 mile = 1.609 kilometers

$$25,000 \text{ in} \times \frac{1 \text{ ft}}{12 \text{ in}} \times \frac{1 \text{ mile}}{5280 \text{ ft}} = .394 \text{ miles}$$

$$.394 \text{ miles} \times \frac{1.609 \text{ km}}{1 \text{ mile}} = .634 \text{ kilometers}$$

Question: One mile on the ground is represented by what distance (in inches) on a 1:25,000 scale map?

Answer:

$$1 \text{ mi} \times \frac{5280 \text{ ft}}{1 \text{ mi}} \times \frac{12 \text{ in}}{1 \text{ ft}} \times \frac{1 \text{ in on map}}{25,000 \text{ in}} = 2.53 \text{ inches on map}$$

Relief Portrayal amd Topographic Maps

A **topographic map** is a two-dimensional representation of three-dimensional land surface. Topographic maps show **relief**—the extent to which an area is flat or hilly. To show land surface, form and vertical relief, topographic maps use contours. A **contour** line connects all points on the map having the same elevation above sea level. Each contour line acts to separate the areas above that elevation from the areas below it (Figure 2).

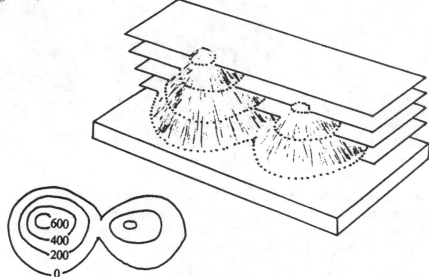

Figure 2

Laboratory Manual for *Conceptual Integrated Science,* © 2007 Addison Wesley

Name _____ Date _____

Earth's Surface—Land and Water: Geologic Cross Sections

Over and Under

Purpose
In this activity, you will construct and interpret maps of geologic cross sections in the subsurface.

Required Equipment and Supplies
protractor
compass
colored pencils and eraser

Discussion
Geologic cross sections show the three-dimensional structure of the subsurface. They provide a "pie slice" of the subsurface and are thus extremely useful for mineral, ore, and oil exploration, We can construct geologic cross sections by using surface information-outcrops of folded and faulted rock sequences, and igneous rock intrusions.

The study of rock deformation is called *structural geology*. When a rock is subjected to compressive stress, it begins to buckle and fold. If the stress overcomes the strength of the rock, the rock exhibits strain and breaks or faults. Structural geology interprets the different types of stress and strain.

We measure the orientation of deformed rock layers using strike and dip (Figure 1). **Strike** is the trend or direction of a horizontal line in an inclined plane. The direction of strike is expressed relative to north. For example, "North × degrees West," or "North × degrees East."

Dip is the vertical angle between the horizontal plane and an inclined plane. Dip is always measured perpendicular to strike. We can think of the direction of dip as the direction a marble would roll down a plane. Hence, we express dip as (1) the direction a marble would roll, and (2) the angle between the inclined and horizontal planes.

Figure 1. Strike and dip. On the rock out-crop, the strike is the line formed from the intersection of a horizontal plane and the tilted rock strata. The dip is the angle between the hori-zontal and tilted strata (plane). The direction of dip is simply the geographical (N, S, E, or W) direction in which a marble would roll down the tilted plane. In the example shown, the outcrop is striking north-west and dipping *45*south-west*.

Geologic map symbols and symbols for strike and dip are shown in Figure 2.

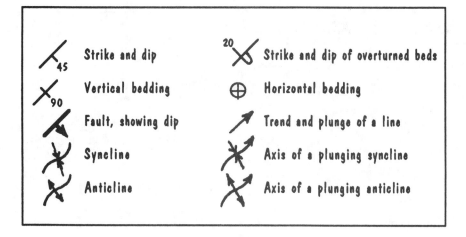

Figure 2. Geologic map symbols.

The orientation of deformed layers can also be obtained through age relationships. Recall that sediments that settle out of water, such as in an ocean or in a bay, are deposited in horizontal layers. The layer at the bottom was deposited first and is therefore the oldest in the sequence of layers. Each new layer is deposited on top of the previous layer. Therefore, in a sequence of sedimentary layers, the oldest layer is at the bottom of the sequence, and the youngest layer is at the top of the sequence. As these sedimentary layers are subjected to stress, they fold and tilt. Each fold has an axis. If the tilted layers dip toward the fold axis, the fold is called a **syncline.** The rocks in the center, or core, are younger than those away from the core. If the tilted layers dip away from the axis, the fold is called an **anticline.** The rocks in the core of the fold of an anticline are older than the rocks away from the core (Figure 3).

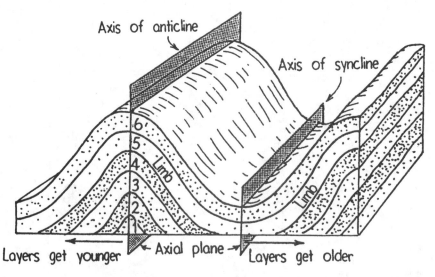

Figure 3. Anticline and syncline folds.

Laboratory Manual for *Conceptual Integrated Science,* © 2007 Addison Wesley

The fold axis itself can be folded, but more often it is simply tilted. Folds in which the axis is tilted are known as *plunging folds* (Figure 4.)

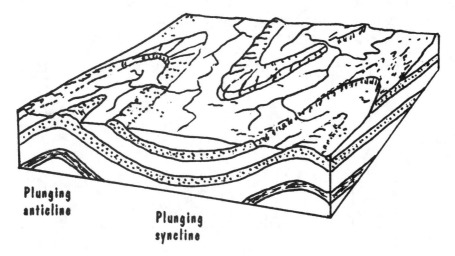

Figure 4. Plunging folds.

When stress exceeds the mechanical strength of a rock, the rock breaks or faults. The type of fault can be deciphered from the ages of the rocks on either side of the fault. Thrust and reverse faults are produced by compression (squeezing) forces. These types of faults have older, structurally deeper rocks pushed on top of younger, structurally higher rocks. Normal faults are produced by tensional (pulling) forces. This type of fault has younger, structurally higher rocks on top of older, structurally deeper rocks. The field evidence for normal faults is repeated layers. The evidence for faults at the earth's surface is "missing" or "repeated" layers of rock.

Shearing forces produce strike-slip faults where the rocks slide past one another. Strike slip faults show no vertical movement, the movement is horizontal. The different types of faults are illustrated in Figure 5.

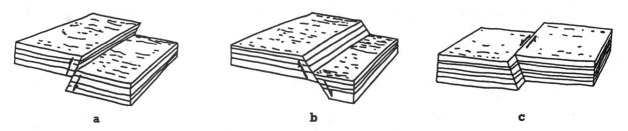

Figure 5. (a) Compression forces produce thrust faults and reverse faults; **(b)** Tensional forces produce normal faults; **(c)** Shearing forces produce strike-slip faults.

Age sequence relationships can also be determined from faults and igneous intrusions. When igneous intrusions or faults cut through other rocks, the intrusion or fault must be younger than the rocks they cut.

In this exercise we will use age relationships, strike and dip orientations, and fault evidence to construct cross sections from map views. A map view only shows the surface features—rock type, sequence of rocks, strike and dip, and fault lines. The cross section view is the subsurface extension of the surface view.

Procedure

Part A: The Construction of a Cross Section Using Strike and Dip Measurements

Step 1: Refer to Figure 6. Transfer locations of contacts and strike and dip symbols to the topographic profile.

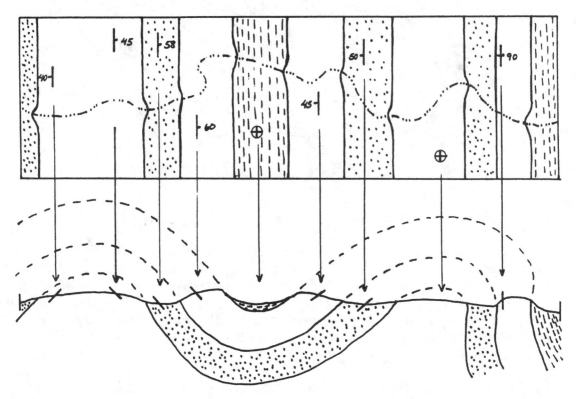

Figure 6

Step 2: Align the protractor to the location of the transferred dip mark on the topographic profile. Depending on the dip direction, rotate the protractor to the measured angle. Mark this angle onto profile. Repeat for all dip marks.

Step 3: Using the dip lines on the topographic profile as guides, draw in the bedding contacts. The lines should be smooth and parallel.

Step 4: Dashed lines may be used to show the eroded surface.

Part B: The Construction of Cross Sections using Age Relationships

Step 1: Determine if there is evidence of folding. Folding will show a mirror image about the fold axis of geologic layers.

Step 2: If the beds are folded, determine if the bedding gets younger or older away from the axis of the fold.

Step 3: If the beds are not folded the disturbed sequence may be due to faulting. Are there repeated layers? Are there missing rock layers? Are the layers offset?

Laboratory Manual for *Conceptual Integrated Science,* © 2007 Addison Wesley

Part C: The Construction of Cross Sections Disturbed by Igneous Intrusions

Step 1: Refer to Figure 7. If the beds are not folded, look to see if there has been a disturbance—faulting or igneous intrusion. Faults and intrusions are always younger than the rock they cut into.

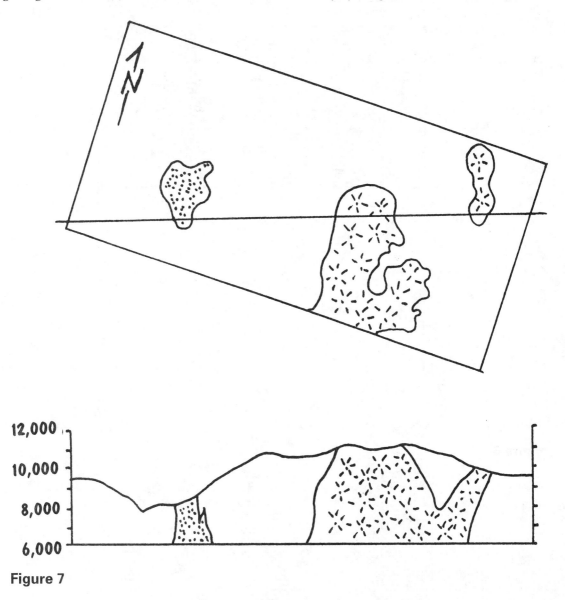

Figure 7

Step 2: Igneous intrusions work their way up from the deep subsurface. Since there is no way to tell the extent of an intrusion in the subsurface, there is a lot of interpretation.

Exercises

1. Use strike and dip symbols to determine the cross section. What is the structure?

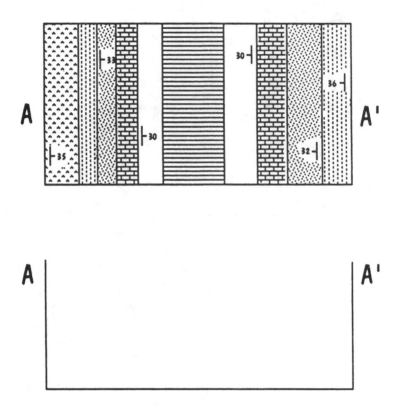

2. Use strike and dip symbols to determine the cross section. What type of structure is this?

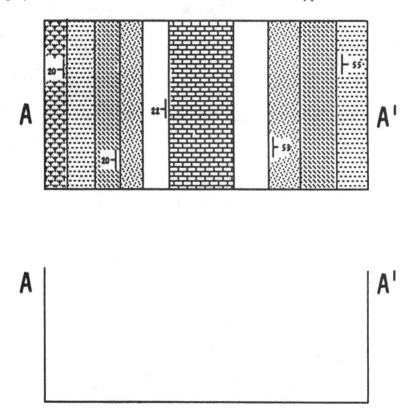

Laboratory Manual for *Conceptual Integrated Science,* © 2007 Addison Wesley

3. Use the relationship and general strike and dip symbols to determine the cross section. What type of structure is this?

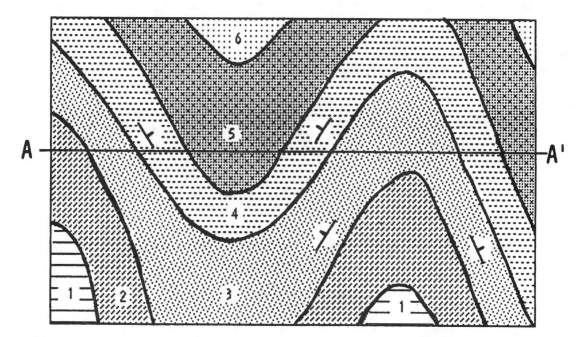

4. This map shows evidence of disturbance and will require some interpretation. Determine the fold structure using the general strike and dip symbols. What type of fold is displayed?

_____

5. The diagonal line represents a fault surface. Show the fault direction movement by placing arrows on the map. What type of fault does this represent?

_____

6. Draw the cross section. What is the oldest structure?

_____

7. What is the youngest structure?

_____

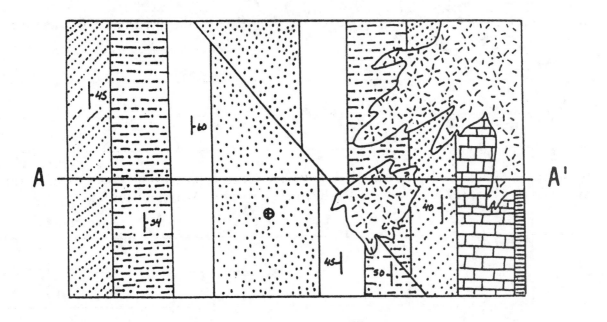

Summing Up

1 What type of structure is displayed in Exercise 1? Is the structure symmetrical or asymmetrical? Can the symmetry be determined from the map view?

_____

_____

_____

2 What type of structure is displayed in Exercise 2? Is the structure symmetrical or asymmetrical? Can the symmetry be determined from the map view?

_____

_____

_____

3 What type of structure is displayed in Exercise 3? What does the dip direction tell us about the structure?

_____

_____

_____

4 What type of fault is displayed in Exercise 4? What evidence supports your answer? What is the fold structure? Based on the drawing, which structure is oldest? Which is youngest?

_____

_____

_____

CONCEPTUAL INTEGRATED SCIENCE | Activity

Earth's Surface—Land and Water: Contour Maps of the Water Table
Walking on Water

Purpose
In this activity, you will construct contour maps of the water table and determine flow paths. Data for map construction can be water table levels obtained from wells and other techniques. You will construct contour lines from data. Your contour lines represent the water table in much the same way as contour lines represent land surface.

Required Equipment and Supplies
ruler
pencil and eraser
protractor

Discussion
Below the surface of continents is an extensive and accessible reservoir of fresh water. This reservoir of subsurface water is divided into two classes—soil moisture and groundwater. Groundwater is water that has percolated into the subsurface and saturated the open pore spaces in the rock or soil. The upper boundary of this saturated zone is called the **water table.** Water in the unsaturated zone above the water tabe is called *soil moisture.* The depth of the water table varies with precipitation and climate.

The water table tends to be a subdued version of the surface topography—sort of an underground surface. Water table contour maps, similar to surface contour maps, show the direction and speed of groundwater flow. This information is extremely useful for water supply management. For instance, simulated water table elevations generated by computer models can be compared with actual water table elevations in order to calibrate and verify the model. A groundwater model will tell you the best location for a well, and the impact that pumping from a new well will have on current water levels. Of particular interest today is the use of groundwater modeling to monitor contaminant transport in the subsurface.

Groundwater flow is influenced by gravitational force. Groundwater flows "downhill" underground, but the path it takes is dependent on *hydraulic head* and not topography. Hydraulic head is higher where the water table is high, such as beneath a hill, and lower where the water table is low, such as beneath a stream valley. So, responding to the force of gravity, water moves from high areas of the water table to low areas of the water table. As groundwater flows, it takes the path of least resistance—the shortest route. For any two lines of equal elevation, the shortest distance between them will be perpendicular to the lines. So, if we think of the path of ground water flow as a *flow line,* the flow line will always be perpendicular to the contour lines.

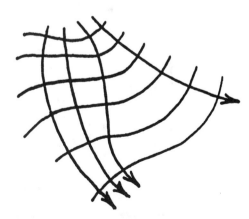

Problem 1: Construction of Water Table Contour Lines

The construction of contour lines for a water table is similar to the construction of surface topographic contour lines. The same procedures and rules for surface contours apply to water table contour maps. The only difference is that you are drawing contour lines of the water table below the ground's surface.

Draw contour lines for the elevation of the water table. Use a contour interval = 10 feet.

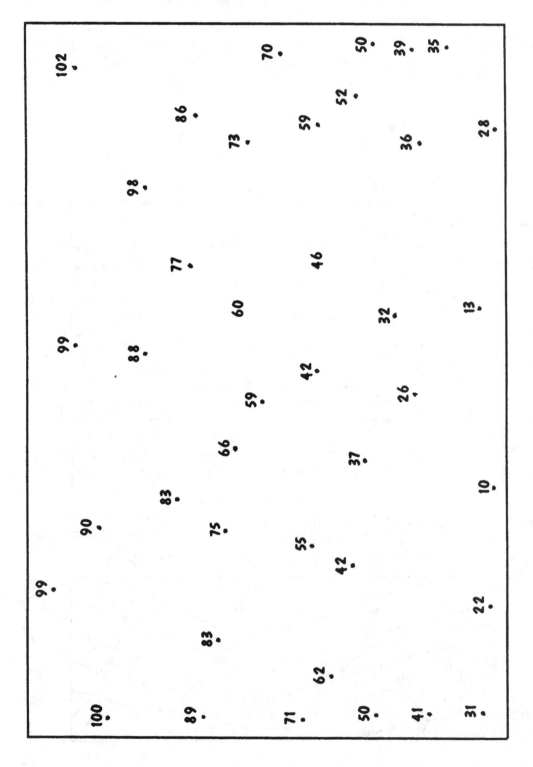

Problem 2: Construction of Flow Lines

Water flows from areas of high hydraulic head to areas of lower hydraulic head. In mapping goundwater flow, the key concept is that the line of flow is perpendicular to the contour lines of the water table. You are provided with the cross section view as a guide to your thinking.

Draw flow lines on the water table contour maps with arrows to show direction. Also draw arrows on the surface of the water table at cross sections to indicate the direction of flow.

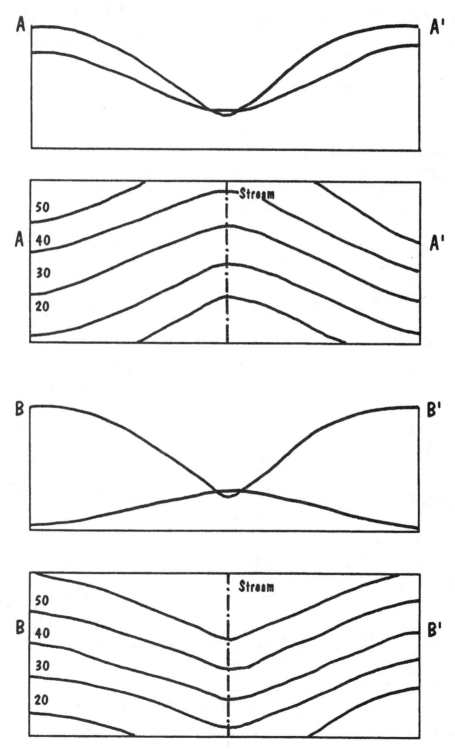

Knowledge of groundwater flow is important for solving problems of groundwater contamination. The most common groundwater contamination comes from sewage—drainage from septic tanks, inadequate or leaking sewer pipes, and farm waste areas. Contamination may also result from landfills and waste dumps where toxic materials and hazardous wastes leach down into the subsurface.

Problem 3: Contaminant Flow Model

Draw lines for the elevation of the water table. Use a contour interval of 10 feet. Water wells and a sewer treatment facility have been plotted on the map. The treatment facility is in disrepair and has developed a leak. Determine which well (if any) will be affected by the flow of contamination from the sewage treatment facility.

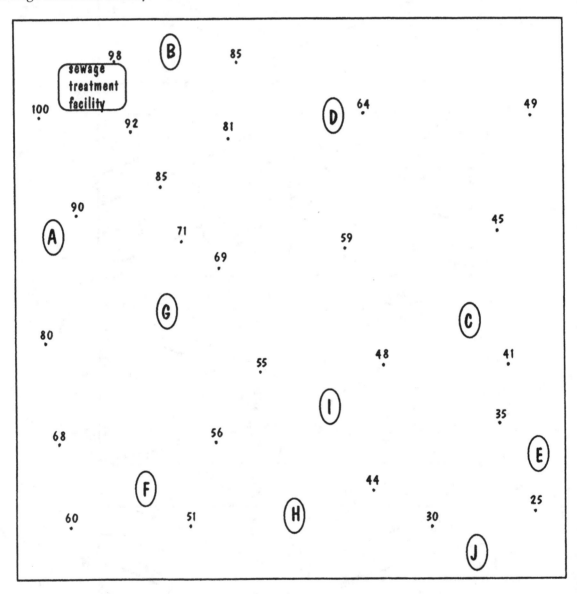

Summing Up

1. In Problem 2, what is the difference between stream A and stream B? In what type of climatic area (dry desert or wet forest) would we likely find stream A? In what type of climatic area would we likely find stream B? Explain.

_____

_____

_____

2. In Problem 3 we are assuming equal rates of pumping for each well. If excessive pumping occurred at well H, would there be a change in the flow of contamination? Which well, if any, would likely become contaminated because of this pumping?

_____

_____

_____

3. In Problem 3 we are assuming that all the soil in the subsurface is homogeneous (the same). If an impermeable clay lens were found at point 55, would the flow of sewage contamination be affected? How about point 35?

_____

_____

_____

Name _____   Date _____

| CONCEPTUAL INTEGRATED SCIENCE | Experiment |

Solar Power I

Purpose
In this experiment, you will measure the Sun's power output by comparison with the power output of a 100-watt lightbulb.

Required Equipment and Supplies
2-centimeter (cm) 6-cm piece of aluminum foil with one side thinly coated with flat black paint
clear tape
meter stick
glass jar with a hole in the metal lid
one-hole stopper to fit the hole in the jar lid
fine-scaled thermometer
clear 100-watt lightbulb with receptacle

Discussion
To measure the power output of a lightbulb is nothing to write home about—but to measure the power output of the sun using only household equipment is a different story. In this equipment we'll do just that—measure to a fair approximation the power output of the Sun. We'll do this by comparing the light from a bulb of known wattage with the light from the Sun, using the simple ratio

$$\frac{\text{Sun's wattage}}{\text{Sun's distance}^2} = \frac{\text{bulb's wattage}}{\text{bulb's distance}^2}$$

We'll need a sunny day to do this experiment.

Procedure
Step 1: With the blackened side facing out, fold the middle of the foil strip around the thermometer bulb, as shown in Figure 1. The ends of the metal strip should line up evenly.

Step 2: Crimp the foil so it completely surrounds the bulb, as in Figure 2. Bend each end of the foil strip outward (Figure 3). On a tabletop or with a meterstick, make a flat, even surface. Use a piece of clear tape to hold the foil to the thermometer.

Step 3: Insert the free end of the thermometer into the one-hole stopper (soapy water or glycerin helps). Remove the lid from the jar, and place the stopper in the lid from the bottom side. Slide the thermometer until the foil strip is located in the middle of the jar. Place the lid on the jar.

Step 4: Position the jar indoors near a window so that sun will shine on it. Prop it at an angle so the blackened side of the foil strip is perpendicular to the rays of the sun. Keep it in this position in the

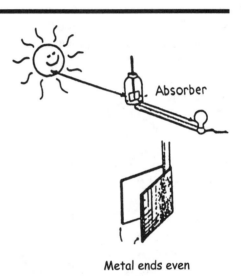

Absorber

Metal ends even

Figure 1

Crimp here

Figure 2

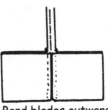
Bend blades outward

Figure 3

sunlight until a stable temperature is reached. If you prefer, do this outside, which means doing Steps 5 and 6 outside.

Stable temperature = _____°C

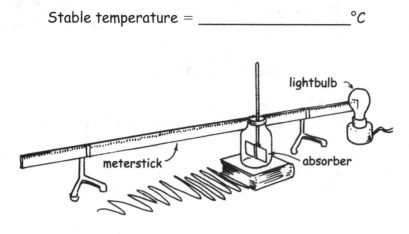

Figure 4

Step 5: Now find the conditions for bringing the foil to the same temperature with a clear 100-watt lightbulb. Set the meter stick on the table. Place the clear 100-watt lightbulb with its filament located at the 0-cm mark of the meter stick (Figure 4). Center the jar at the 95-cm mark with the blackened side of the foil strip perpendicular to the light rays from the bulb. You may need to put some books under the jar to align it properly.

Step 6: Turn the lightbulb on. Slowly move the jar toward the lightbulb, 5 cm at a time, allowing the thermometer temperature to stabilize each time. As the temperature approaches the reading reached in Step 4, move the jar only 1 cm at a time. When the bulb maintains the same temperature obtained from the sun for about 2 minutes, turn the lightbulb off.

Step 7: Measure as exactly as possible the distance in meters between the foil and the filament of the bulb. Record the distance.

Distance from light filament to foil strip = _____m

Step 8: Because you know the Sun's distance from the thermometer in meters is 1.5×10^{11} m, and you know the distance of the lightbulb from the thermometer, and wattage of the bulb, you know three of the four values for the ratio equality stated earlier. With simple rearrangement

$$\text{Sun's wattage} = \frac{(\text{bulb's wattage})(\text{Sun's distance})^2}{(\text{bulb's distance})^2}$$

Show your work:

Sun's wattage = _____W

Step 9: Use the Sun's wattage to compute the number of 100-watt lightbulbs needed to equal the Sun's power. Show your work.

Number of 100-watt lightbulbs = _____

        Laboratory Manual for *Conceptual Integrated Science,* © 2007 Addison Wesley

Summing Up

The accepted value for the Sun's power is 3.8×10^{26} W. How close was your experimental value? List factors you can think of that affect the difference.

_____

_____

CONCEPTUAL INTEGRATED SCIENCE	Activity

Weather: Clouds and Precipitation

Indoor Clouds

Purpose
In this activity, you will illustrate the formation of a cloud from the condensation of water droplets.

Required Equipment and Supplies
large glass jar
measuring cup for water
small metal baking tray
tray of ice cubes

Discussion
Clouds are made up of millions of tiny water droplets and/or ice crystals. Cloud formation takes place as rising moist air expands and cools.

As the Sun warms the Earth's surface, water is evaporated from the oceans, lakes, steams, and rivers. The process of evaporation changes the water molecules from a liquid phase to a vapor phase. Because evaporation is greater over warmer waters than cooler waters, tropical locations have a higher water vapor content than polar locations. Water vapor content is dependent on temperature. For any given temperature, there is a limit to the amount of water vapor in the air. This limit is called the dew point. When this limit is reached, the air is saturated. A measure of the amount of water vapor in the air is called *humidity* (the mass of water per volume of air). The amount of water vapor in the air varies with geographic location and may change according to temperature.

Warm air can hold more water vapor before becoming saturated than can cooler air. When the air temperature falls, the air becomes saturated—the amount of water vapor the air can hold reaches its limit. As the air cools below the dew point, the water vapor molecules condense onto the nearest available surface. Condensation is the change from a vapor phase to a liquid phase. The condensation of water vapor molecules on small airborne particles produces cloud droplets, which in turn become clouds.

Most clouds form as air rises, expands, and cools. There are several reasons for the development of clouds. When the temperatures of certain areas of the Earth's surface increase more readily than other areas, air may rise from thermal convection. As this warm air rises, it mixes with the cooler air above and eventually cools to its saturation point. The moisture from the warm air condenses to form a cloud. As the cloud grows, it shades the ground from the sun. This cuts off the surface heating and upward thermal convection—the cloud dissipates. After the cloud is gone, the ground again heats up to start another cycle of thermal convection.

Clouds also form as a result of topography. Air rises as it moves over mountains. As it rises, it cools. If the air is humid, clouds form. As the air moves down the other side of the mountain, it warms. This air is drier because most of the moisture has been removed to form the clouds on the other side. It is therefore more common to find cloud formation on the windward side of mountains rather than on the leeward side.

Clouds may form as a result of converging air. When cold air moves into a warm air mass, the warm air is forced upward. As it rises, it cools, and water vapor condenses to form clouds. Cold fronts are associated with extensive cloudiness and thunderstorms. When warm air moves into a cold air mass, the less dense warmer air rides up and over the colder, denser air. Warm fronts result in widespread cloudiness and light precipitation that may extend for thousands of square kilometers.

Procedure

Step 1: Fill a glass jar with about 1 inch of very hot water. Do not use boiling water—the glass may break.

Step 2: Place ice cubes in a metal baking tray and set the tray on top of the jar. Make sure there is a good seal.

Step 3: Observe a "cloud." As the ice above cools the air inside the top of the jar, water vapor condenses into water droplets to form a "cloud."

Summing Up

1. How is your formation of clouds the same as the formation of clouds in the sky?

2. How is it different?

3. Why are coastal tropical areas more humid than desert areas?

4. Why do we believe that warm air rises?

5. How does topography contribute to the formation of desert areas? For example, what role do the Sierra Nevada mountains have in keeping the Nevada side of the mountain a desert?

CONCEPTUAL INTEGRATED SCIENCE	Activity

Earth's History: Geologic Time

Geologic Time Scale and Relative Dating

Purpose

In this activity, you will use the Principles of Relative Dating and the Geologic Time Scale to investigate relative dating of rock layers.

Required Equipment and Supplies

Conceptual Integrated Science text, pencil and eraser

Discussion

Earth's history is recorded in the rocks of its crust. We cannot see the processes that created rocks that are millions and even billions of years old, but we can observe present-day rock-forming processes and their results. If the resulting present-day characteristics are similar to the characteristics of the older rocks in question, we can infer that the processes that operated in the past are the same as those that operate today. This is the **Principle of Uniformitarianism**—the physical, chemical, and biological laws that operated in the distant past continue to operate today. Simply stated, the present is the key to the past.

Geologists of long ago observed how present-day sediment is deposited horizontally, one layer at a time. This basic observation, called **original horizontality,** is one of the fundamental principles of **relative dating**—the process of determining the order in which geological events occurred. The principles of relative dating, as the name suggests, help us determine the chronological order of geological events, but not the actual dates of these events.

The five key principles of relative dating are as follows:

1. **Original horizontality:** Sediment layers are deposited evenly, with each new layer laid down nearly horizontally over older rocks and sediment.

2. **Superposition:** In an undeformed, layered sequence of rocks, each layer is older than the one above and younger than the one below. In other words, the rock record was formed from the bottom layer to the top.

3. **Cross-cutting relationships:** Any type of igneous intrusion or fault that cuts through a rock body or layer is younger than the rock through which it cuts.

4. **Inclusions:** Any rock inclusion is older than the rock containing it.

5. **Faunal succession:** The evolution of life is recorded in the rock record in the form of fossils. Fossil organisms follow one another in a definite, irreversible sequence. Once an organism becomes extinct, it is never again seen at later times in the fossil record.

Problem 1: Relative Time—What Came First?

The cross-section below shows the results of several geologic events. To the right of the diagram, list the sequences of geologic history starting with the oldest event and ending with the youngest event.

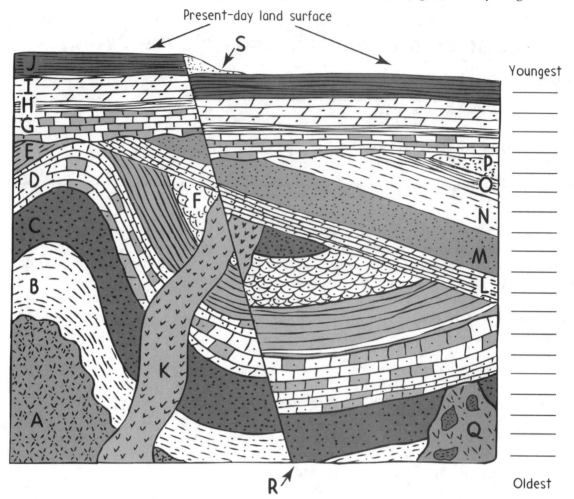

Present-day land surface

Youngest

Oldest

Summing Up

1. What geological events happened before layer "L" was deposited? [Hint: Think tectonics.]

_____

_____

_____

_____

2. What type of event does feature "Q" represent? Which principles help you to approximate the age range of "Q"?

_____

_____

_____

_____

3. Feature "R" is a fault. What is the approximate age range of "R"? What type of fault is "R"? What type of force generated the fault, compression or tension? Support your answer.

_____

_____

_____

_____

Problem 2: Life and the Geologic Time Scale

The Geologic Time Scale was developed through the use of relative dating, and fossils played a large role. Through meticulous analysis of fossils, geologists were able to arrange different groups of fossils—and the rock layers in which they were found—into a chronological sequence.

Because each time period can be recognized by the fossils it contains, the fossils found in rocks can be used to identify other rocks of the same age in other regions of the earth. Actual specific dates were later applied to the geologic time scale with radiometric dating (a process that measures the ratio of radioactive isotopes to their decay product).

Use Section 26.3 of your text to match the life form with the appropriate geologic time period.

Geologic Time Period		Life Form
Quaternary	_____	A) Trilobite
Tertiary	_____	B) Flowering plants
Cretaceous	_____	C) Age of humans
Jurassic	_____	D) Swampy environments
Triassic	_____	E) Emergence of dinosaurs
Permian	_____	F) Age of mammals
Carboniferous	_____	G) True pines and redwoods
Devonian	_____	H) First reptiles
Silurian	_____	I) Age of fishes
Ordovician	_____	J) Emergence of land plants
Cambrian	_____	K) First fish

CONCEPTUAL INTEGRATED SCIENCE	Activity

Earth's History: The Rock Record

Reading the Rock Record

Purpose

The *principle of superposition* can be used to determine the relative ages of rock layers at a single location. Such locations, where rocks are exposed at the Earth's surface, are called *outcrops*. For regions with several outcrops, geologists also use the *principle of lateral continuity* to determine relative ages. The principle of lateral continuity states that the layer-forming sediments we now see in rock outcrops were originally deposited and solidified into contiguous horizontal layers. When fully formed, such layers extended continuously for great distances until an obstruction was encountered. In this activity, you will reconstruct a simple geological history of a hypothetical area by examining diagrams of several rock outcrops from that area. You will then use fossils and their respective periods of existence in geologic time to determine the ages of the rock layers.

Required Equipment and Supplies

Conceptual Integrated Science text, pencil and eraser

Discussion

Perhaps the world's most spectacular display of a rock record is the Grand Canyon of the Colorado River in Arizona. The many layers of sedimentary rock exposed in the canyon walls and the thicknesses of the different rock layers are testimony to great geologic activity over millions of years. The vertical sequence of its cake-like layering provides a great example for understanding relative dating. And the roughly horizontal rock layers that often stretch continuously for hundreds of miles are wonderful examples of lateral continuity.

But even the Grand Canyon does not exhibit a complete and continuous geological history. In fact, even though some rock layers may have been deposited without interruption, nowhere on Earth do we find a continuous sequence of rock from Earth's formation to the present time. Weathering and erosion, crustal uplifts, and other geologic processes interrupt and/or prevent the deposition of a continuous sequence. These time gaps, or breaks, in the rock record are called **unconformities.** An unconformity between two rock layers can represent one or more rock layers that used to be present but were eroded away before deposition of the overlying layer. Or an unconformity can represent a long period of time in which deposition did not occur. As such, an unconformity indicates a significant time gap between the underlying and overlying layers. Unconformities are discovered by carefully observing the relationships of rock layers and fossils.

Correlation of Rock Outcrops

Piecing together Earth's history and conducting other geological investigations often involves matching up, or correlating, rock outcrops that are separated geographically by large distances. In contrast to the Grand Canyon, most rock layers are not horizontally continuous from one rock outcrop to the next. The processes that contribute to unconformities—faulting, folding, uplift, and weathering and erosion—also act to break the lateral continuity of layers. And younger, unconsolidated sediment and soil end up burying older rock layers so that all that remains visible are geographically separated rock outcrops. Because of such occurrences, it is often difficult to tell within a study area if similar-looking rock layers at different outcrops used to be part of the same, once-continuous layer.

To correlate one isolated rock outcrop to another involves looking for similar features among the different outcrops. We begin with the physical characteristics of the rocks—color, mineralogy, and grain size, for example. If the various rock layers in the isolated outcrops have characteristics that are similar enough, and they have consistent vertical sequences, we can safely assume that the layers were at one time continuous.

For example, in Outcrop A (Figure 1) we might have a limestone layer sandwiched between a black shale layer on top and a red sandstone layer on the bottom. Then, if we find a similar sequence of rock layers at Outcrops B and C, the three geographically separated outcrops are correlated and were once laterally continuous layers.

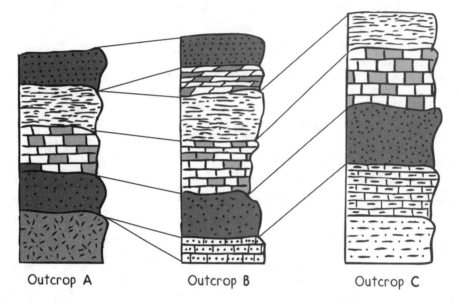

Outcrop A   Outcrop B   Outcrop C

Figure 1. Correlation of Outcrops A, B, and C shows that the sandstone, limestone, and shale layers are in the same relative vertical sequence—evidence that all three Outcrops were, at one time, laterally continuous. Parallel lines are drawn between the Outcrops for the layers on which the correlation is based. Note the "pinched" lines from Outcrop B to Outcrop A. These "pinched" lines show that two rock layers are missing from Outcrop A. The boundaries between the two uppermost and the two lowermost rock layers in Outcrop A are unconformities. The missing layers indicate two time gaps in the rock record at Outcrop A.

The Importance of Fossils
The job of correlating rock layers between outcrops is made easier when fossils are found. Recall the principle of faunal succession—that the evolution of life is recorded in the form of fossils, and that fossil organisms follow one another in a definite, irreversible sequence in time. When fossils are found, they can be used as distinctive markers in sequences of sedimentary rocks. They not only aid in correlating rock layers, they can also be used as age indicators of the rock.

A group of fossil organisms that lived during the same range of time and are found in one or more layers is called a **fossil assemblage**. So if our limestone layer at outcrop A contains an assemblage of certain marine-type fossils, and then the same assemblage is found at outcrops B and C, we have even stronger evidence to support our correlation that the rock layers were at one time continuous. Further investigation into the type of fossils in the assemblages allows us to determine the time period(s) during which the rocks were formed.

It is important to note that the presence of a similar or even identical fossil assemblage from rocks in two different locations does not mean that those rocks are exactly the same age. Correlation of fossil assemblages tells us that the ages of rock layers at different outcrops are similar in that they were formed within the same time range. Even though there may be an uncertainty of thousands to millions of years, such an approximation is sufficient for many purposes. (Remember, Earth is 4.5 *billion* years old!)

Construction of a Stratigraphic Column

The first geologic time scale was constructed by using the principle of superposition and the correlation of fossil assemblages in sedimentary rocks from numerous regions. Primitive fossils were placed at the bottom of the time scale column not because of their primitive nature but because of superposition—they were found at the lowest layers in several (or many) different sequences of rock layers. The assumption was that rocks—and the fossils in them—found at the bottom of a sequence of rock layers are older than those found at the top.

The following exercise will provide insight into the methods that were used to construct the geologic time scale. Geologists show the vertical chronological succession of sedimentary rock layers from oldest at the bottom to youngest at the top by first constructing a diagram called a **stratigraphic column.**

Procedure

Four stratigraphic columns are depicted in Figure 2. The columnar stratigraphic sections are from outcrops in four widely separated, hypothetical regions. The rock layers in these outcrops were at one time continuous between all four regions. Various episodes of uplift and erosion have broken the continuity these rock layers used to share.

Please note: P is the oldest layer.

Step 1: To see which layers match up, draw lines connecting the boundaries of similar rock units from one outcrop to the next (see the example in Figure 1). In the individual stratigraphic columns, which rock layers are missing between the various regions? Note the missing layers for each column.

Step 2: In the blank column to the left, construct a new stratigraphic column based on the correlation between the four different regions. The new column will include every rock layer found at the four outcrops. Draw the different rock layers, or write the letter for each layer, in their proper order— oldest at the bottom to youngest at the top. Once again, remember that *P* is the oldest layer.

Going Further

Fossils document the evolution of life through time. Certain fossils that are geographically widespread and limited to a short span of geologic time are called index fossils. Index fossils can be used for correlating rocks of the same age and determining the approximate ages of rock layers. When a specific index fossil cannot be found, geologists look for groups of fossils—a fossil assemblage. In many ways, an assemblage of fossils can provide even more precise information. The more data one has, the clearer the picture becomes.

Now back to our hypothetical rock outcrops. The different layers of rock contain different fossil assemblages. Your task now is to assign time periods to the layers.

Step 3: Figure 3 shows the age ranges for fossils in our hypothetical study area. Many fossils overlap one another in time. Given that all rock units are for the Paleozoic Era, assign time periods to the rock layers in Figure 2. Write the names of the periods next to the respective rock layers in your stratigraphic column from Step 2.

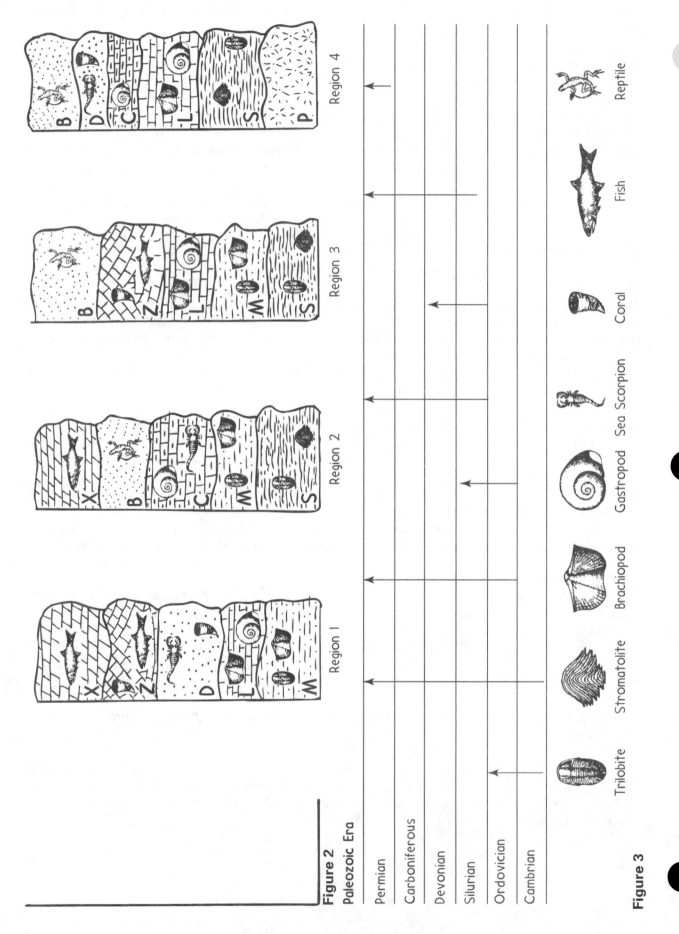

Figure 2
Paleozoic Era

Permian

Carboniferous

Devonian

Silurian

Ordovician

Cambrian

Region 1 Region 2 Region 3 Region 4

Trilobite Stromatolite Brachiopod Gastropod Sea Scorpion Coral Fish Reptile

Figure 3

Laboratory Manual for *Conceptual Integrated Science,* © 2007 Addison Wesley

Summing Up

1. List the missing rock layers for each region.

_____

_____

2. The appearance of identical fossils in each of the sections facilitates the correlation of rock layers between each outcrop. The reptile in three of the outcrops is not present in Region 1. What could this indicate? Support your answer.

_____

_____

3. Our particular fish fossil is found in Layers Z and X and has a time range extending from the Devonian to the Permian. How can we assign Layer X a specific time period? Explain.

_____

_____

4. There are several rock layers that have more than one designated time period. Name these layers. How can fossils extending over more than one time period be found in one rock layer?

_____

_____

5. There are many different types of fossils in our hypothetical rock layers. What do the majority of these fossils have in common? (Need a hint? Think habitat and environment.)

_____

_____

CONCEPTUAL INTEGRATED SCIENCE	Experiment

The Solar System: The Sun

Sunballs

Purpose
In this experiment, you will estimate the diameter of the Sun.

Required Equipment and Supplies
small piece of cardboard
meterstick

Discussion
Take notice of the round spots of light on the shady ground beneath trees. These are sunballs—images of the Sun. They are cast by openings between leaves in the trees that act as pinholes. The diameter of a sunball depends on its distance from the small opening that produces it. Large sunballs, several centimeters or so in diameter, are cast by openings that are relatively high above the ground, while small ones are produced by closer "pinholes." The interesting point is that the ratio of the diameter of the sunball to its distance from the pinhole is the same as the ratio of the Sun's diameter to its distance from the pinhole.

Because the Sun is approximately 150,000,000 km from the pinhole, careful measurement of this ratio tells us the diameter of the Sun. That's what this experiment is all about. Instead of finding sunballs under the canopy of trees, you'll make your own easier-to-measure sunballs.

Procedure
Poke a small hole in a piece of cardboard with a pen or sharp pencil. Hold the cardboard in the sunlight and note the circular image that is cast on a convenient screen of any kind. This is an image of the Sun. Unless you're holding the card too close, note that the solar image size does not depend on the size of the hole in the cardboard (pinhole), but only on its distance from the pinhole to the screen. The greater the distance between the image and the cardboard, the larger the sunball.

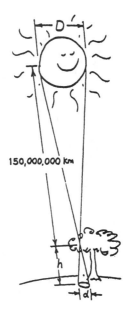

150,000,000 km

Position the cardboard so the image exactly covers a dime, or something that can be accurately measured. Carefully measure the distance to the small hole in the cardboard. Record your measurements as a ratio:

$$\frac{\text{diameter of dime}}{\text{distance from dime to pinhole}} = \underline{\hspace{4cm}}$$

Because this is the same ratio as the diameter of the sun to its distance, then

$$\frac{\text{diameter of dime}}{\text{distance from dime to pinhole}} = \frac{\text{diameter of Sun}}{\text{distance from Sun to pinhole}}$$

Which means you can now calculate the diameter of the Sun!

Diameter of the Sun = _____

Summing Up

1. Will the sunball still be round if the pinhole is square shaped? Triangular shaped? (Experiment and see!)

_____

_____

_____

2. If the Sun is low so the sunball is elliptical, should you measure the short or the long width of the ellipse for the sunball diameter in your calculation of the Sun's diameter? Why?

_____

_____

_____

3. If the Sun is partially eclipsed, what will be the shape of the sunball?

_____

_____

_____

WHAT SHAPE Do SUNBALLS HAVE DURING A PARTIAL ECLIPSE OF THE SUN ?

Name _____   Date _____

The Solar System: Overview of the Solar System

Ellipses

Purpose
In this activity, you will investigate the geometry of the ellipse.

Required Equipment and Supplies
about 20 centimeters (cm) of string
2 thumbtacks
pencil and paper
a flat surface that will accept thumbtack punctures

Discussion
An ellipse is an oval-like closed curve defined as the locus of all points about a pair of foci (focal points), the sum of whose distances from both foci is a constant. Planets orbit the sun in elliptical paths, with the Sun's center at one focus. The other focus is a point in space, typified by nothing in particular.

Elliptical trajectories are not confined to "outer space." Toss a ball and that parabolic path it seems to trace is actually a small segment of an ellipse. If extended, its path would continue through Earth and swing about Earth's center and return to its starting point. In this case, the far focus is Earth's center. The near focus is not typified by anything in particular. The ellipse is very stretched out, with its long axis considerably longer than its short axis. We say the ellipse is very *eccentric*. Toss the rock faster and the ellipse is wider—less eccentric. The far focus is still Earth's center, and the near focus is nearer to Earth's center than before. Toss the rock at 8 km/s and both foci will coincide at Earth's center. The elliptical path is now a circle—a special case of an ellipse. Toss the rock faster, and it follows an ellipse external to Earth. Now we see the near focus is the center of Earth and the *far* focus is beyond—again, at no particular place. As speed increases and the ellipse becomes more eccentric again, the far focus is *outside* Earth's interior.

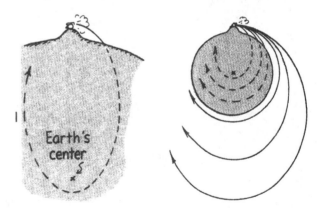

Constructing an ellipse with pencil, paper, string, and tacks is interesting. Let's do it!

Procedure
Place a loop of string around two tacks or pushpins and pull the string taut with a pencil. Then slide the pencil along the string, keeping it taut. (To avoid twisting of the string, it helps if you make the top and bottom half in two separate operations.)

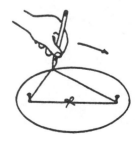

Label each focus of your ellipse (the location of the pushpins). Repeat, using different focus separation distances, and you'll notice that the greater the distance between foci, the more eccentric will be the ellipse.

Construct a circle by bringing both foci together. A circle is a special case of an ellipse.

Another special case of an ellipse is a straight line. Determine the positions of the foci that give you a straight line.

The shadow of a ball face on is a circle. But when the shadow is not face on, the shape is an ellipse. The photo below shows a ball illuminated with three light sources. Interestingly enough, the ball meets the table at the focus for all three ellipses.

Summing Up

1. Which of your drawings approximates Earth's orbit about the Sun?

_____

2. Which of your drawings approximates the orbit of Halley's Comet about the Sun?

_____

3. What is your evidence for the definition of the ellipse: that the sum of the distances from the foci is constant?

_____

_____

_____

CONCEPTUAL INTEGRATED SCIENCE	**Experiment**

The Solar System: Overview of the Solar System

Reckoning Latitude

Purpose

In this experiment, you will build and use the necessary instruments for determining how far above the equator you are on this planet. This will be done by measuring the angle of Polaris in the sky. Actual measurements will need to be made outside of class because measurements can only be performed at nighttime.

Required Material and Supplies

protractor
pencil with unused eraser
pushpin
drinking straw
plumb line and bob
tape

Discussion

Latitude is a measure of your angular distance north or south of the equator (Figure 1). To determine your latitude in the Northern Hemisphere, it is convenient to use Polaris, the North Star, as a guide. At the North Pole you would find that the North Star is directly above you—90° above the horizon. This is also your angular distance from the equator (Figure 2a). If you were to travel closer to the equator, the North Star would recede behind you to some lower angle above the horizon, which would always match your angular distance from the equator (Figure 2b). If, for example, you measured the North Star to be 45° above the horizon, you would be at an angular distance of 45° north of the equator. At the equator you'd find the North Star to be 0° above the horizon (Figure 2c). (It turns out that refraction at this grazing angle complicates viewing.)

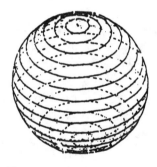

Figure 1. Latitudes are parallel to the equator.

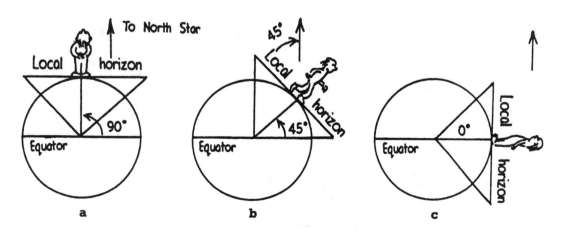

Figure 2. The angle at which the North Star appears above the horizon is also your angular distance from the equator.

Procedure

An astrolabe is a simple instrument for measuring angles above the horizon (altitude). To construct an astrolabe, tape a straw to the straight edge of a protractor as shown in Figure 3. Place the eraser end of a pencil against the hole in the protractor and insert the pushpin through the other side of the hole into the eraser—the pencil serves as a convenient handle. Next, tie a plumb bob to the needle (a string attached to a weight).

To measure the altitude of any object viewed through the straw, have a partner read the angle the plumb bob line makes with the protractor. Practice the use of your astrolabe by measuring the angle of various objects around you. Work in pairs or small groups so that one person can read the angle while the other is sighting. Take turns looking through the astrolabe to be sure of your readings. You may want to make several measurements and find the average. The angle that you read should be between 0° and 90°. Be sure that this angle is not the number of degrees from the zenith (straight up). You can convert such readings into degrees from the horizon (straight out) by subtracting the number of degrees from 90°. Challenge yourself or others to estimate measurements without using the astrolabe. How accurate are these "naked-eye" estimates?

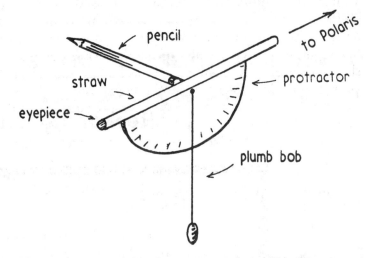

Figure 3. Astrolabe for measureing altitude.

Measuring the Altitude of Polaris

The North Star is fairly easy to locate because of its position relative to the Big and Little Dippers, two well-known constellations that stand out in the northern sky. To find the North Star from the Big Dipper, draw an imaginary line between the last two stars of the Big Dipper's pan. Follow this line away from the top of the Big Dipper and the first bright star you cross is the North Star (Figure 4). Note that the North Star is the first star of the Little Dipper's handle.

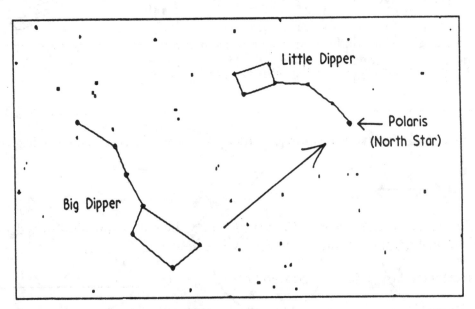

Figure 4. Polaris, the North Star, can be found in the northern sky from its position relative to the Big and Little Dippers.

Laboratory Manual for *Conceptual Integrated Science,* © 2007 Addison Wesley

To find your latitude, you need simply measure the altitude of the North Star above the horizon. It so happens that the North Star is not *exactly* above the North Celestial Pole, so for a more accurate determination of your latitude, a minor but significant correction may be necessary. The North Star lies about 3/4 of a degree away from the North Celestial Pole in the direction of Cassiopeia, a "W"-shaped constellation (Figure 5). Hence, the altitude of the North Star may not correspond exactly to your latitude. If you see Cassiopeia much "above" the North Star (higher up from the horizon), then subtract 3/4 of a degree from your measured altitude of Polaris to obtain your latitude. If, on the other hand, Cassiopeia is "beneath" the North Star, add 3/4 of a degree. If Cassiopeia is anywhere to the left or right of the North Star, then the correction is not necessary.

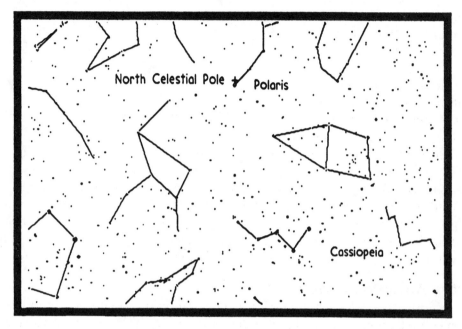

Figure 5. Polaris is offset from the pole toward Cassiopeia.

Summing Up

1. What was your measured altitude of Polaris?

2. According to this measurement, at what latitude are you located?

3. By how many degrees does your measured latitude differ from the accepted latitude given by your instructor?

4. One degree of latitude is equivalent to 107 km (66 miles). By how many kilometers does your determined latitude differ from the accepted value in kilometers?

    (Your Determined Latitude − Accepted Latitude) × 107 = _____

5. How could the astrolabe be used to provide evidence that the world is round?

CONCEPTUAL INTEGRATED SCIENCE	Activity

The Solar System: The Inner Planets

Tracking Mars

Purpose

In this activity, you will plot the orbit of Mars. You will employ the technique of Johannes Kepler and use data obtained by Tycho Brahe four centuries ago.

Required Equipment and Supplies

four sheets of plain grid graph paper   compass
protractor   ruler
sharp pencil

Discussion

In the early 1500s, the Polish astronomer Nicolaus Copernicus used observations and geometry to determine the orbital radius and orbital period for the planets known at the time. Copernicus based his work on the revolutionary assumption that the planets moved around the Sun. In the late 1500s, before telescopes were invented, the Danish astronomer Tycho Brahe made 20 years worth of extensive and accurate measurements of planets and bright stars. Near the end of Tycho's career, he hired a young German mathematician, Johannes Kepler, who was assigned the task of plotting the orbit of Mars using Tycho's data.

Kepler started by drawing a circle to represent Earth's orbit (not bad, considering Earth's orbit only approximates a circle, which he didn't know at the time). Since Mars takes 687 Earth days to orbit the Sun once, Kepler paid attention to observations that were exactly 687 days apart. In this way, Mars would be in the same place while Earth would be in a different location. Two angular readings of Mars' location from the same location on Earth 687 days apart was all that was needed—where the two lines crossed was a point on the orbit of Mars. Plotting many such points did not trace out a circle, as Kepler had expected—rather, the path was an ellipse. Kepler was the first to discover that if the planets orbit the Sun, they did so in elliptical rather than circular paths. He then went on to plot a better orbit for Earth based on observations of the Sun, and further refined the plotted orbit of Mars.

In this activity, you will duplicate the work of Kepler, simplifying somewhat by assuming a circular orbit for Earth—it turns out the difference is minor, and the elliptical path of Mars is evident. From Brahe's extensive tables, we'll use only data shown in Table 1.

Procedure

Step 1: You'll want a sheet of graph paper with about a 14 × 14-inch working area. If need be, tape two legal-size sheets of borderless graph paper together, or four regular 8 1/2″ × 11″ sheets.

Step 2: Make a dot at the center of your paper to represent the Sun. Place a compass there, and draw a 10-cm radius circle to represent the orbit of Earth around the Sun. Draw a light line from the center to the right of the paper, and mark the intersection with Earth's orbit 0°. This is the position of Earth on March 21. All your plotting will be counterclockwise around the circle from this reference point.

Table 1. Brahe's data are grouped in 14 pairs of Mars' sightings. For the first nine pairs, the first line of data is for Mars in opposition—when Sun, Earth, and Mars were on the same line—when Mars was 90° to Earth's horizon, directly overhead at midnight. The second line of data are positions measured 687 days later, when Mars was again in the same place in its orbit, and Earth in a different place, where a different angle was then measured. All angles given in the table read from a 0° reference line—the line from the Sun to Earth at the vernal equinox, March 21. Mars at points 10–14 are nonopposition sightings. The first line of Point 10, for example, shows that when Earth was at 277°, Mars was seen not directly overhead, but at 208.5° with respect to the 0° reference line. Then 687 days later, Earth was at 235°, where Mars was seen at 272.5°. The data are neatly arranged for plotting—something that took Kepler years to do.

Mars Orbit Point	Mo	Date Day	Year	Earth Position (Ecliptic)	Mars Position (Ecliptic)
1	11	28	1580	66.5	66.5
	10	16	1582	22.5	107
2	1	7	1583	107	107
	11	24	1584	62.5	144
3	2	10	1585	141.5	141.5
	12	29	1586	97.5	177
4	3	16	1587	175.5	175.5
	1	31	1589	132	212
5	4	24	1589	214.5	214.5
	3	12	1591	171.5	253.5
6	6	18	1591	266.5	266.5
	5	5	1593	225	311
7	9	5	1593	342.5	342.5
	7	24	1595	300.5	29.5
8	11	10	1595	47.5	47.5
	9	27	1597	4	90
9	12	24	1597	92.5	92.5
	11	11	1599	48	130.5
10	6	29	1589	277	208.5
	5	16	1591	235	272.5
11	8	1	1591	308	260.5
	6	18	1593	266.5	335
12	9	9	1591	345.5	273
	7	27	1593	304	347.5
13	10	3	1593	9.5	337.5
	8	20	1595	327	44.5
14	11	23	1593	60.5	350.5
	10	10	1595	17	56

Step 3: Locate the first point in Mars' orbit, Point 1, from Table 1. Do this by first marking with a protractor the position of Earth along the circle for the date November 28, 1580. This is 66.5° above the 0° reference line. Draw a dot to show Earth's position at this time.

Step 4: Mars at this time was in *opposition*—opposite to the Sun in the sky. A line from the Sun to Earth at this time extends radially outward to Mars. Draw a line from the center of your circle (the Sun) to Earth's position at this time, and beyond Earth through Mars. Where is Mars along this line? You'll need another sighting of Mars 687 days later when Mars is at the same place and Earth is in another.

Step 5: If you were to add 687 days to November 28, 1580, you'd get October 16, 1582. At that date, Mars was measured to be lower in the sky—actually 107.0° with respect to the 0° reference line of March 21. With respect to the reference line, use a protractor and ruler and draw a line at 107.0° as shown in Figure 1. Where your two lines intersect is a point along Mars' orbit.

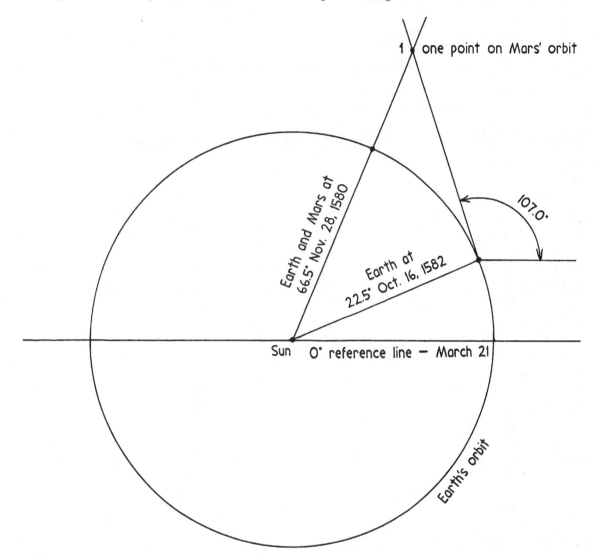

Figure 1. Your plot should look like this for finding Mars Orbit Point 1; data are in the first pair of lines of Table 1. Angles from Earth to Mars are given with respect to the 0° reference line of March 21.

Step 6: With care, plot the 13 other intersections that represent points along Mars' orbit, using the data in Table 1.

Step 7: Connect your points, either very carefully by freehand, or with a French curve.

Bravo—you have plotted the orbit of Mars using Kepler's method from four centuries ago!

Summing Up

1. Place the point of your compass on the position of the Sun and place the pencil on the first point of the plotted orbit of Mars. Draw a circular orbit with that radius. Does that circle match the orbit of Mars? If not, describe the differences.

_____

2. Does your plot agree with Kepler's finding that the orbit is an ellipse? _____

_____

3. During what month are the orbits of Mars and Earth closest? _____

Appendix A

Significant Figures and Uncertainty in Measurement

Units of Measurement

All measurements consist of a number and a unit. Both are necessary. If you say that a friend is going to give you 5, you are telling only *how many*. You also need to tell *what*—five fingers, five cents, five dollars, or five corny jokes. If your instructor asks you to measure the length of a piece of wood, saying that the answer is 26 is not correct. She or he needs to know whether the length is 26 centimeters, feet, or meters. All measurements must be expressed using a number and an appropriate unit.

Numbers

Two kinds of numbers are used in science—those that are counted or defined, and those that are measured. There is a great difference between a counted or defined number and a measured number. The exact value of a counted or defined number can be stated, but the exact value of a measured number cannot.

For example, you can count the number of chairs in your classroom, the number of fingers on your hand, or the number of quarters in your pocket with absolute certainty. Counted numbers are not subject to error (unless you counted wrong).

Defined numbers are about exact relations and are defined to be true. The defined number of centimeters in a meter, the defined number of seconds in an hour, and the defined number of sides on a square are examples. Defined numbers also are not subject to error (unless you forget the definition).

Every measured number, no matter how carefully measured, has some degree of uncertainty. What is the width of your desk? Is it 89.5 centimeters, 89.52 centimeters, 89.520 centimeters, or 89.5201 centimeters? You cannot state its exact measurement with absolute certainty.

Uncertainty in Measurement

Uncertainty (or margin of error) in a measurement can be illustrated by the two different metersticks in Figure A. The measurements are of the length of a tabletop. Assuming that the zero end of the meterstick has been carefully and accurately positioned at the left end of the table, how long is the table?

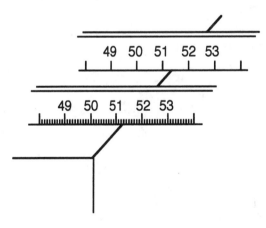

The upper scale in the figure is marked off in centimeter intervals. Using this scale, you can say with certainty that the length is between 51 and 52 centimeters. You can say further that it is closer to 51 centimeters than to 52 centimeters; you can estimate it to be 51.2 centimeters.

The lower scale has more subdivisions and has a greater precision because it is marked off in millimeters. With this meterstick, you can say that the length is definitely between 51.2 and 51.3 centimeters, and you can estimate it to be 51.25 centimeters.

Note how both readings contain some digits that are known, and one digit (the last one) that is estimated. Note also that the uncertainty in the reading of the lower meterstick is less than that of the top meterstick. The lower meterstick can give a reading to the hundredths place, and the top meterstick to the tenths place. The lower meterstick is more *precise* than the top one. So, digits tell us the magnitude of a measurement while the location of the decimal point tells us the precision.

Significant Figures

Significant figures are the digits in any measurement that are known with certainty plus the one digit that is estimated and hence is uncertain. The measurement 51.2 centimeters (made with the top meterstick in Figure A) has three significant figures, and the measurement 51.25 centimeters (made with the lower meterstick) has four significant figures. The right-most digit is always an estimated digit. Only one estimated digit is ever recorded as part of a measurement. It would be incorrect to report that the length of the table (Figure A) is 51.253 centimeters as measured with the lower meterstick. This five-significant-figure value would have two estimated digits (the 5 and 3) and would be incorrect, because it indicates a *precision* greater than the meterstick can obtain.

Standard rules have been developed for writing and using significant figures, both in measurements and in values calculated from measurements.

Rule 1
In numbers that do not contain zeros, all the digits are significant.

> EXAMPLES:
> | | |
> |---|---|
> | 4.1327 | five significant figures |
> | 5.14 | three significant figures |
> | 369 | three significant figures |

Rule 2
All zeros between significant digits are significant.

> EXAMPLES:
> | | |
> |---|---|
> | 8.052 | four significant figures |
> | 7059 | four significant figures |
> | 306 | three significant figures |

Rule 3
Zeros to the left of the first nonzero digit serve only to fix the position of the decimal point and are not significant.

> EXAMPLES:
> | | |
> |---|---|
> | 0.0068 | two significant figures |
> | 0.0427 | three significant figures |
> | 0.0003506 | four significant figures |

Rule 4
In a number with digits to the right of the decimal point, zeros to the right of the last nonzero digit are significant.

> EXAMPLES:
> | | |
> |---|---|
> | 53 | two significant figures |
> | 53.0 | three significant figures |
> | 53.00 | four significant figures |
> | 0.00200 | three significant figures |
> | 0.70050 | five significant figures |

Laboratory Manual for *Conceptual Integrated Science*, © 2007 Addison Wesley

Rule 5

In a number that has no decimal point and that ends in one or more zeros (such as 3600), the zeros that end the number may or may not be significant.

The number is ambiguous in terms of significant figures. Before the number of significant figures can be specified, further information is needed about how the number was obtained. If it is a measured number, the zeros are not significant. If the number is a defined or counted number, all the digits are significant.

Confusion is avoided when numbers are expressed in scientific notation. All digits are taken to be significant when expressed this way.

EXAMPLES:

4.6×10^{-5}	two significant figures
4.60×10^{-5}	three significant figures
4.600×10^{-5}	four significant figures
2×10^{-5}	one significant figures
3.0×10^{-5}	two significant figures
4.00×10^{-5}	three significant figures

Rounding

Calculators often display eight or more digits. How do you round such a display to, say, three significant figures? Three rules govern the process of deleting unwanted (insignificant) digits from a calculator number.

Rule 1

If the first digit to the right of the last significant figure is less than 5, that digit and all the digits that follow it are simply dropped.

EXAMPLES:
51.234 rounded to three significant figures becomes 51.2.

Rule 2

If the first digit to be dropped is a digit greater than 5, or if it is a 5 followed by a digit other than zero, the excess digits are dropped, and the last retained digit is increased in value by one unit.

EXAMPLES:
51.35, 51.359, and 51.3598 rounded to three significant figures all become 51.4.

Rule 3

If the first digit to be dropped is a 5 not followed by any other digit, or if it is a 5 followed only by zeros, an odd–even rule is applied.

That is, if the last retained digit is even, its value is not changed, and the 5 and any zeros that follow are dropped. But if the last digit is odd, its value is increased by one. The intention of this odd–even rule is to average the effects of rounding off.

EXAMPLES:
74.2500 to three significant figures becomes 74.2.
89.3500 to three significant figures becomes 89.4.

Significant Figures and Calculated Quantities

Suppose that you measure the mass of a small wooden block to be 2 grams on a balance, and you find that its volume is 3 cubic centimeters by dipping it beneath the surface of water in a graduated cylinder. The density of the piece of wood is its mass divided by its volume. If you divide 2 by 3 on your calculator, the reading on the display is 0.6666666. However, it would be incorrect to report that the density of the block of wood is 0.6666666 gram per cubic centimeter. To do so would be claiming a higher degree of precision than is warranted. Your answer should be rounded off to a sensible number of significant figures.

The number of significant figures allowable in a calculated result depends on the number of significant figures in the measured data, and on the type of mathematical operation(s) used in calculating. There are separate rules for multiplication and division, and for addition and subtraction.

Multiplication and Division

For multiplication and division, an answer must have the number of significant figures found in the number with the fewest significant figures. For the density example given above, the answer must be rounded off to one significant figure, 0.7 gram per cubic centimeter. If the mass were measured to be 2.0 grams, and if the volume were still taken to be 3 cubic centimeters, then the answer must still be rounded off to 0.7 gram per cubic centimeter. If the mass were measured to be 2.0 and the volume 3.0 or 3.00 cubic centimeters, the answer must be rounded off to two significant figures: 0.67 gram per cubic centimeter.

Study the following examples. Assume that the numbers being multiplied or divided are measured numbers.

EXAMPLE A:

8.536 × 0.47 = 4.01192 (calculator answer)

The input with the fewest significant figures is 0.47, which has two significant figures. Therefore, the calculator answer 4.01192 must be rounded off to 4.0.

EXAMPLE B:

3840 ÷ 285.3 = 13.45916 (calculator answer)

The input with the fewest significant figures is 3940, which has three significant figures. Therefore, the calculator answer 13.45916 must be rounded off to 13.5.

EXAMPLE C:

36.00 ÷ 3.000 = 12 (calculator answer)

Both inputs contain four significant figures. Therefore, the correct answer must also contain four significant figures, and the calculator answer 12 must be written as 12.00. In this case, the calculator gave too few significant figures.

Addition and Subtraction

For addition and subtraction, the answer should not have digits beyond the last digit position common to all the numbers being added or subtracted. Study the following examples:

Laboratory Manual for *Conceptual Integrated Science*, © 2007 Addison Wesley

EXAMPLE A:

```
    34.6
    18.8
+   15
    68.4 (calculator answer)
```

The last digit position common to all numbers is that just to the left of where a decimal point is placed or might be placed. Therefore, the calculator answer of 68.4 must be rounded off to 68.

EXAMPLE B:

```
    20.02
    20.002
+   20.0002
    60.0222 (calculator answer)
```

The last digit position common to all numbers is the hundredths place. Therefore, the calculator answer of 60.0222 must be rounded off to 60.02.

EXAMPLE C:

```
    345.56
−   245.5
    100.06 (calculator answer)
```

The last digit position common to both numbers in this subtraction is the tenths place. Therefore, the answer must be rounded off to 100.1.

Percentage Error

A measured value is best compared to an accepted value by the percentage of difference rather than the size of the difference. Measuring the length of a 10-centimeter pencil to +/− one centimeter is quite a bit different from measuring the length of a 100-meter track to the same +/− centimeter. The measurement of the pencil shows a relative uncertainty of 10%. The track measurement is uncertain by only 1 part in 10,000, or 0.01%.

The relative uncertainty or relative margin of error in measurements, when expressed as a percentage, is the *percentage of error*. It tells by what percentage a quantity differs from a known accepted value as determined by skilled observers using high-precision equipment. It is a measure of the *accuracy* of the method of measurement, which includes the skill of the person making the measurement. The percentage of error is equal to the difference between the measured value and the accepted value of a quantity divided by the accepted value, and then multiplied by 100. (Note: % error should always be presented as a positive number. Take the absolute value if necessary.)

$$\% \text{ error} = \frac{(\text{accepted value} - \text{measured value})}{(\text{accepted value})} \times 100$$

For example, suppose that the measured value of the acceleration of gravity is found to be 9.44 m/s². The accepted value is 9.81 m/s². The percentage error is

$$\% \text{ error} = \frac{(9.81 \text{ m/s}^2 - 9.44 \text{ m/s}^2)}{(9.81 \text{ m/s}^2)} \times 100 = 3.77\%$$

Appendix B

Graphing

Many quantities are dependent on one another. The circumference of a circle, for example, depends on the circle's diameter, and vice versa. Tables, equations, and graphs show how dependent quantities are related. Investigating relationships between dependent quantities make up much of the work of physical science. Tables, equations, and graphs are important physical science tools.

Tables

Tables give values of dependent variables in list form. Table A, for example, lists the instantaneous speed of an object relative to its elapsed time of motion. A table organizes experimental data and helps the investigator to deduce relationships.

Table A

Elapsed Time (seconds)	Instantaneous Speed (meters/second)
0	0
1	10
2	20
3	30
4	40
.	.
.	.
.	.
t	$10t$

The relationship between two or more dependent variables can be described using words or can be more concisely expressed using an *equation*. For example, to say that the speed of free fall depends on acceleration and time is said concisely by the equation $v = v_0 + gt$, where v_0 is the initial speed when time $t = 0$. When free fall begins from $v_0 = 0$, we have just $v = gt$.

A *graph* is a pictorial representation of the relationship between dependent variables, such as those found in an equation. By looking at the shape of a graph, we can quickly tell a lot about how the variables are related. For this reason, graphs can help clarify the meaning of an equation or table of numbers. Also, a graph can help reveal the relationship between variables when the equation is not already known. Experimental data are often graphed for this reason.

The most common and simplest graph is the Cartesian graph, where values of one variable are represented on the vertical axis, called the y-axis, and values of the other variable are represented on the horizontal or x-axis. If variable y is *directly proportional* to variable x then the curve will appear to rise from left to right (Figure A). Immediately you can tell that as x increases, y also increases. If, however, x is *inversely proportional* to y then the curve will appear to decrease from left to right (Figure B). In this case, as x increases, y decreases.

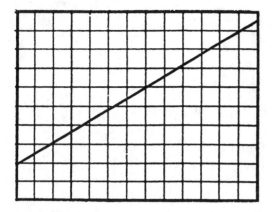

Figure A

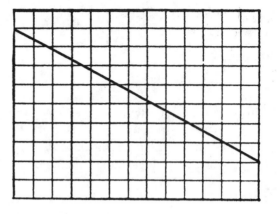

Figure B

Direct and inverse proportionalities are types of linear relationships. Linear relationships have straight-line graphs—the easiest kind to interpret. Figure C shows a graph of the equation $v = gt$. Speed v is plotted along the y-axis, and time t along the x-axis. As you can see, there is a linear relationship between v and t. Here v increases in direct proportion to t; double t and v doubles, triple t and v triples, and so on.

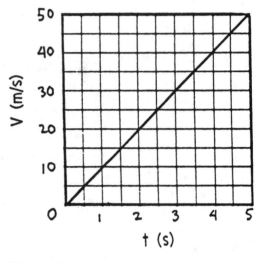

Figure C

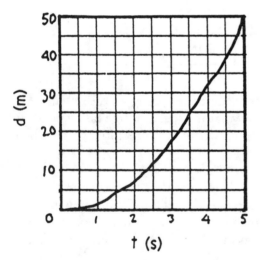

Figure D

Figure D shows a graph of distance d compared to time t in the equation $d = 1/2\ gt^2$. We see by the non-linear curve that d is not directly proportional to t. Interestingly enough, if we make a graph of d versus t^2, a straight line results. This is because distance d is directly proportional to t^2, with a slope that is calculated to be 5 m/s^2, or 1/2 g (try it yourself and see). Plotting various powers of nonlinear quantities until they form a straight line is one way to find the equation that relates the quantities.

Laboratory Manual for *Conceptual Integrated Science*, © 2007 Addison Wesley

Area Under the Curve

An important feature of a graph that often has physical significance is the area under the curve. Consider the rectangular area bounded by the graph in Figure E. Here the area is the product of the y and x dimensions, speed v and time t, respectively (the area of any rectangle is the product of two perpendicular sides). In this case, the area vt has physical significance, for $vt = d$, the distance traveled during the time interval t. Because 50 m/s × 5 s = 250 m, we see the distance traveled in 5 s is 250 m.

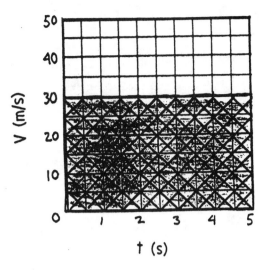

Figure E

The area need not be rectangular. The area beneath any curve of v vs. t represents the distance traveled. Similarly, the area beneath a curve of acceleration vs. time gives the velocity acquired ($v = at$), or the area beneath a force vs. time curve gives the momentum acquired ($Ft = mv$). (What does the area beneath a force versus distance curve give?) The area under various curves, including rather complicated ones, can be found by way of an important branch of mathematics—*integral calculus*.

Graphing with *Conceptual Integrated Science*

Graphs clarify equations by making them pictorial. You will develop some rudimentary graphing skills in your physical science laboratory. Of particular worth is the lab activity that utilizes a ranging device and a computer, "Sonic Ranger."

Questions*

Figure F is a graphical representation of the velocity of a ball dropped from a cliff.

1. How long did the ball take to hit the bottom?

2. What was its velocity when it struck bottom?

3. What does the decreasing slope of the graph tell you about the acceleration of the ball with increasing speed?

4. Did the ball reach terminal velocity before hitting the bottom? If so, about how many seconds were required for it to reach its terminal velocity?

5. What is the approximate height of the cliff?

6. How sudden was the impact?

---

*Answers: 1) 9 s; 2) 25 m/s; 3) Acceleration decreases as velocity increases (because of air resistance); 4) Yes (because slope curves to zero), about 6 s; 5) Falling distance is nearly 180 m [the area under the curve is about 71 squares—each square represents 2.5 m (5 m/s × 0.5 s = 2.5 m)]; 6) The impact was very sudden as evidenced by the rapid decrease in velocity at the 9th second (what would the graph look like if the ball bounced back up?).

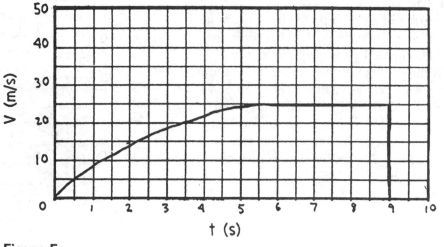

Figure F

Plotting a Graph

The first step in plotting experimental data on a graph is to choose the dimension and scale of the x- and y-axes. Larger graphs allow for greater precision. For this reason, the axes should be drawn and labeled so that the data are spaced out over as much of the graph paper as possible. The axes should also be labeled with the units they represent.

Data from a data table are entered into the graph as a series of points. Each point represents the magnitudes—as specified by the axes—of both variable quantities (Figure G). If there is a relation between the variables, then as the data are being plotted, a pattern emerges. In Figure G we see the pattern is a straight line.

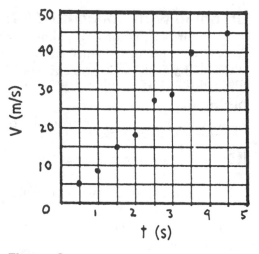

Figure G

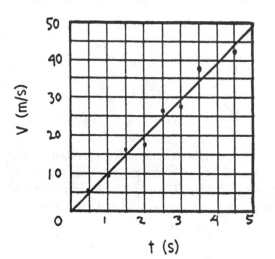

Figure H

Due to experimental error, which includes the limited precision of instruments, data points often deviate from the apparent pattern. For example, data points showing the pattern of a straight line may not fall exactly in a straight line. In this case, a single straight line that approximates all of the data points may be drawn using a ruler or straight edge (Figure H). This straight line need not touch any of the data points for it represents the best fit of all the points together. Nonlinear relationships can similarly be sketched with a curve that best fits all the data points.

The *slope* of a linear relationship is found by measuring how high the line on the graph rises compared to how far out it extends. This may be done by drawing a right triangle that has the sloping line as its hypotenuse (Figure I). The slope is equal to the height of the triangle (its "rise") divided by the base (its "run").

$$\text{Slope} = \frac{\text{rise}}{\text{run}}$$

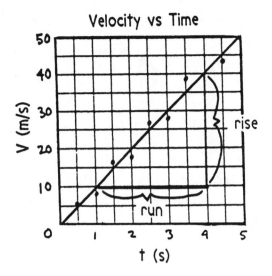

Figure 1

Rise is read as the height the right triangle rises relative to the *y*-axis, and run is read as the distance the triangle runs relative to the *x*-axis. The units of the *x*- and *y*-axes must be included. In Figure I, for example, the rise equals 30 m/s, and the run equals 3 s. The slope of the line, therefore, equals:

$$\text{slope} = \frac{\text{rise}}{\text{run}} = \frac{30 \text{ m/s}}{3s} = 10 \frac{\text{m/s}}{\text{s}} = 10 \frac{\text{m}}{s^2}$$

The units show us what the slope represents. In this case, the sloped line represents the acceleration (m/s^2) of the object. Note the distinction between the *angle* of the line, 45°, and the *slope* of the line, 10 m/s^2.

The final step to creating a graph is assigning a brief but descriptive title, such as "Velocity vs. Time" (Figure 1).

Appendix C

Conversion Factors

In science, it is often necessary to convert from one type of unit into another. Units are converted by multiplying the given quantity by a *conversion factor*. A conversion factor is a ratio derived from an equality, hence, a conversion factor always equals 1. Consider the following example:

$$12 \text{ inches} = 1 \text{ foot}$$

A given quantity divided by the same quantity always equals one.

$$\frac{12 \text{ inches}}{12 \text{ inches}} = 1 \qquad \text{or} \qquad \frac{1 \text{ foot}}{1 \text{ foot}} = 1$$

Because 12 inches and 1 foot represent the same quantity, we can perform the following substitutions:

$$\frac{12 \text{ inches}}{1 \text{ foot}} = 1 \qquad \text{or} \qquad \frac{1 \text{ foot}}{12 \text{ inches}} = 1$$

The above two ratios are conversion factors. Because conversion factors are equal to 1, multiplying a quantity by a conversion factor does not change that quantity. While the quantity *doesn't* change, the units *do* change. For example, suppose some item was measured at 60 inches in length. This quantity may be converted into feet by multiplying by the appropriate conversion factor:

$$(60 \text{ inches}) \frac{(1 \text{ foot})}{(12 \text{ inches})} = 5 \text{ feet}$$

$$\Uparrow \qquad\qquad \Uparrow \qquad\qquad \Uparrow$$

quantity   conversion   quantity
in inches   factor   in feet

Note how the units of inches cancel each other because they are found in both the numerator and denominator. Similarly, the length of an item measured in feet can be converted to inches using the reciprocal conversion factor:

$$(5 \text{ feet}) \frac{(12 \text{ inches})}{(1 \text{ foot})} = 60 \text{ inches}$$

$$\Uparrow \qquad\qquad \Uparrow \qquad\qquad \Uparrow$$

quantity   conversion   quantity
in feet   factor   in inches

To derive a conversion factor, consult a table of unit equalities, such as Table A. The unit you want to obtain in your answer should be set in the numerator of the conversion factor, and the unit you want to cancel should be set in the denominator.

Several conversion factors may be linked together when the relationship between two units is not given directly. For example, suppose you are to convert 5.00 inches into kilometers. In this case, you might convert inches into centimeters and then centimeters into meters followed by meters into kilometers:

$$(5.00 \text{ inches}) \frac{(2.54 \text{ cm})}{(1 \text{ inch})} \frac{(1 \text{ meter})}{(100 \text{ cm})} \frac{(1 \text{ km})}{(1000 \text{ meters})} = 0.000127 \text{ kilometers}$$

Alternatively, you might convert inches into feet and then feet into miles followed by miles into kilometers:

$$(5.00 \text{ inches}) \frac{(1 \text{ foot})}{(12 \text{ inch})} \frac{(1 \text{ mile})}{(5280 \text{ feet})} \frac{(1.609 \text{ km})}{(1 \text{ mile})} = 0.000127 \text{ kilometers}$$

The technique of following your units guides you in selecting the proper conversion factor. For example, suppose we wish to convert 15.0 calories per minute into joules per second, which are the units of watts. Then

$$(15.0 \frac{cal}{min}) \frac{(4.187 \text{ J})}{(1 \text{ cal})} \frac{(1 \text{ min})}{(60 \text{ sec})} = 1.22 \frac{J}{sec} = 1.22 \text{ watts}$$

Similarly, we may convert 55.0 miles per hour into meters per second:

$$(55.0 \frac{mi}{hr}) \frac{(5280 \text{ ft})}{(1 \text{ mi})} \frac{(12 \text{ in})}{(1 \text{ ft})} \frac{(2.54 \text{ cm})}{(1 \text{ in})} \frac{(1 \text{ m})}{(100 \text{ cm})} \frac{(1 \text{ hr})}{(3600 \text{ sec})} = 24.6 \frac{m}{sec}$$

In all cases, unwanted units cancel and only wanted units remain.

Table A. Units of measurement

UNITS OF LENGTH		EXACT ?*
1 kilometer (km)	= 1000 meters (m)	yes
1 meter (m)	= 1000 millimeters (mm)	yes
	= 100 centimeters (cm)	yes
1 micron (u)	= 1×10^{-6} m	yes
1 nanometer (nm)	= 1×10^{-9} m	yes
1 angstrom (A)	= 1×10^{-10} m	yes
1 inch (in)	= 2.54 cm	yes
1 mile (mi)	= 1.6093440 km	yes
	= 5280 feet	yes
UNITS OF MASS AND WEIGHT		
1 kilogram (kg)	= 1000 g	yes
1 gram (g)	= 1000 milligrams (mg)	yes
1 kg	= 2.205 pounds	no
1 pound (lb)	= 453.6 g	no
	= 16 ounces (oz)	yes
	= 4.448 newtons (N)	no
UNITS OF VOLUME		
1 liter (L)	= 1000 milliliters (mL)	yes
1 mL	= 1 cubic centimeter (cc or cm^3)	yes
1 L	= 1.057 quarts (qt)	no
1 cubic inch (in^3)	= 16.39 mL	no
1 gallon (gal)	= 4 quarts (qt) = 8 pints (pt)	yes
UNITS OF ENERGY		
1 calorie (cal)	= 4.184 joules (J)	yes
	= amount of heat needed to raise 1 g of water by 1 °C	yes
1 kilocalorie (kcal)	= 1000 cal = 1 nutritional calorie (C)	yes
	= 4.184 kJ	yes
1 joule (J)	= 1×10^7 erg = 1 kg m^2/s^2	yes
1 electron volt (eV)	= 24.217 kcal/mol = 1.602×10^{-19} J	no
1 kilowatt hour (kWh)	= 3.60×10^6 J	no
UNITS OF PRESSURE		
1 atmosphere (atm)	= 760 mm Hg = 760 torr	yes
	= 14.70 pounds per square inch (psi)	no
	= 29.29 in. Hg	no

*Conversion factors involving exact relationships contain an unlimited number of significant figures (see Appendix A).

Laboratory Manual for *Conceptual Integrated Science,* © 2007 Addison Wesley

Exercises:

1. Convert 20.0 calories into joules.

(20.0 calories)
⇑   ⇑   ⇑
quantity   conversion   quantity
in calories   factor   in joules

2. Convert 26.2 miles into kilometers.

(26.2 miles)
⇑   ⇑   ⇑
quantity   conversion   quantity
in miles   factor   in kilometers

3. Convert 2.00 miles into centimeters.

Answers:

1. Use the conversion factor 4.187 joules/1 calorie to convert 20.0 calories into 83.7 joules.

2. Use the conversion factor 1.609 km/1 mile to convert 26.2 miles into 42.2 kilometers.

3. Convert miles into feet and then feet into inches followed by inches into centimeters to get the answer 322,000 centimeters. (See Appendix A for rules concerning significant figures).